DATE DUE

11-28-94	
11-24 97	
8-10-06	
UPI 261-2505	PRINTED IN U.S.A.

Genetics and Heredity

Genetics and Heredity

The Blueprints of Life

TORSTAR BOOKS
New York • Toronto

TORSTAR BOOKS
41 Madison Avenue
Suite 2900
New York, NY 10010

THE HUMAN BODY
Genetics and Heredity:
The Blueprints of Life

Publisher
Bruce Marshall

Art Director
John Bigg

Creation Coordinator
Harold Bull

Editor
John Clark

Managing Editor
Ruth Binney

Commissioning Editor
Hal Robinson

Contributors
Arthur Boylston, Julian Chomet, Loraine Fergusson, Jack Singer

Text Editors
Wendy Allen, Mike Darton, Martyn Page, Sandy Shepherd

Researchers
Maria Pal, Jazz Wilson

Picture Researchers
Jan Croot, Kate Duffy, Dee Robinson

Layout and Visualization
Eric Drewery, Ted McCausland

Artists
Mick Gillah, Nicki Kemball, Aziz Khan, Mick Saunders

Cover Design
Moonink Communications

Cover Art
Paul Giovanopoulos

Production Director
Barry Baker

Production Coordinator
Janice Storr

Business Coordinator
Candy Lee

Planning Assistant
Avril Essery

International Sales
Barbara Anderson

Marshall Editions, an editorial group that specializes in the design and publication of scientific subjects for the general reader, prepared this book. Marshall has written and illustrated standard works on technology, animal behavior, computer usage and the tropical rain forests which are recommended for schools and libraries as well as for popular reference.

Series Consultants

Donald M. Engelman is Professor of Molecular Biophysics and Biochemistry and Professor of Biology at Yale. He has pioneered new methods for understanding cell membranes and ribosomes, and has also worked on the problem of atherosclerosis. He has published widely in professional and lay journals and lectured at many universities and international conferences. He is also involved with National Advisory Groups concerned with Molecular Biology, Cancer, and the operation of National Laboratory Facilities.

Stanley Joel Reiser is Professor of Humanities and Technology in Health Care at the University of Texas Health Science Center in Houston. He is the author of *Medicine and the Reign of Technology* and coeditor of *Ethics in Medicine: Historical Perspectives and Contemporary Concerns;* and coeditor of the anthology *The Machine at the Bedside.*

Harold C. Slavkin, Professor of Biochemistry at the University of Southern California, directs the Graduate Program in Craniofacial Biology and also serves as Chief of the Laboratory for Developmental Biology in the University's Gerontology Center. His research on the genetic basis of congenital defects of the head and neck has been widely published.

Lewis Thomas is Chancellor of the Memorial Sloan-Kettering Cancer Center in New York City and University Professor at the State University of New York, Stony Brook. A member of the National Academy of Sciences, Dr. Thomas has served on advisory councils of the National Institutes of Health.

Consultants for Genetics and Heredity

William A. Horton is an Associate Professor of Pediatrics and Medicine at The University of Texas Medical School at Houston and the Director of the Medical Genetics Program at that Institution. He is a Diplomate of the American Board of Medical Genetics and belongs to many professional societies. He has published widely and has contributed to many books in the area of medical genetics. His major research interest lies in the inherited disorders of connective tissues.

Patricia D. Murphy is affiliated with the Clinical Cytogenics Laboratory and the Department of Human Genetics at Yale University, New Haven, Connecticut. She has published research articles spanning the fields of immunology, somatic and molecular genetics. Currently, her research interest is linkage analysis in humans.

Medical Advisor
Arthur Boylston

Library of Congress
Cataloging in Publication Data
Main entry under title:

Genetics and Heredity: The Blueprints of Life

Includes index.
1. Human genetics. I. Torstar Books (Firm).
QH431.G4323 1985 573.2'1 85–20685
ISBN 0–920269–65–6 (Genetics & Heredity)
ISBN 0–920269–22–2 (The Human Body Series)
ISBN 0-920269–66–4 (leatherbound)
ISBN 0–920269–67–2 (school ed.)

20 19 18 17 16 15 14 13 12 11
10 9 8 7 6 5 4 3

Printed in Belgium

Contents

Introduction:

The Human Inheritance

Like truth, breeding will out. Certain physical characteristics undoubtedly "run in the family." Even a newborn baby is declared by admiring relatives to have "Dad's eyes" and "Mom's mouth" or other particular features that are somehow different from the average yet specific to the parents. The Hapsburgs, rulers of the Holy Roman Empire from the 1200s, became celebrated through paintings of the family over several centuries which revealed them all to have the characteristic "Hapsburg nose." Large feet, a prominent chin, a short neck, an unusual pigmentation of the eyes — all of these and many other characteristics are the results of heredity, of genetic programming involving both parents.

How the process of heredity actually works has been learned only during this century, yet the theoretical groundwork was proposed in the 1860s by Gregor Mendel, who was way ahead of his time. This obscure Moravian monk worked out the mathematical nature of inheritance through experimentation with plants, and he spoke of "factors" that made it work. It was the turn of the century before scientists took up the challenge to identify those factors — the "working units" of heredity.

The chromosome had been discovered in the nuclei of cells in the late 1870s (by the German anatomist Walther Flemming), and in 1909 the Danish botanist Wilhelm Johannsen coined the word "gene." The genes were the carriers of inherited characteristics, stacked together in rows along the chromosomes — an association worked out by the pioneering American geneticist Thomas Hunt Morgan. A knowledge of the chemical composition of genes had to wait for the birth of molecular biology, when the structural analysis and then isolation of RNA and DNA helped to crack the genetic code. The scientific mechanism of heredity was known at last, paving the way for modern genetic engineering. Yet, were we to ask them, children might request from their parents no more than to inherit the family silver.

Families in formal pose at a wedding reveal inherited resemblances. An old saying advises a bridegroom to look at the bride's mother to see what the bride will be like in later life. In this family the bride's mother, standing on the right, has passed on her facial features to her daughter, but the bride appears to have inherited her hair color from her father.

Chapter 1

Ancestral Legacies

Acres of corn stretch out to the horizon ready to be harvested by a squadron of reapers. The worldwide production and yield of crops such as this have increased a thousandfold though the genetic manipulation of various strains and species. Similar genetic engineering techniques have been applied to experimental animals, and perhaps one day may be used to improve the health of humankind.

An awareness of heredity — the passing on of inherited characteristics from one generation to the next — began the moment that people became aware that like begat like, that humans produced humans, dogs gave birth to dogs, and corn produced yet more corn. Although this awareness goes back almost as long as man has inhabited the Earth, it is only in the last hundred years that science has been able to explain not only how this happens but also why offspring have a combination of the characteristics of their parents.

This knowledge has had remarkable effects. It has, for example, allowed the breeding of specific characteristics in crop plants, so that farmers may cultivate more effectively. The result is that today they produce more than three thousand pounds of corn per acre, compared with medieval farmers who could accomplish only one-tenth as much. And animals have been bred so that they produce more meat, milk or wool. For humans, knowledge of heredity has unraveled the cause of a wide variety of afflictions, some of which can now be prevented.

George Bernard Shaw once wrote: "Take care to get born well." He could well have written instead: "Hope that you receive a good set of inherited characteristics from your parents — for you are stuck with them for life."

The information within our cells that causes us to develop in a particular way is stored in genes, specific parts of chromosomes which are present at some time in every body cell. But it is only in the twentieth century that the thin chromosome strands within cells have been associated with the process of heredity. Since this remarkable discovery, the science of genetics — the study of heredity and variation, and of the resemblances and differences between organisms — has really blossomed.

Although a great deal is known to science about how chromosomes and their genes exert their effects, ideas about heredity can be traced back at

least six thousand years to stone carvings from Chaldea which describe different pedigrees, showing the inheritance of particular characteristics of a horse's mane. Other carvings made by the ancient Babylonians show the cross-pollination of date palm trees by the application of pollen from the male palm to the pistils of flowers on the female tree in order to produce fruits.

Some time later, at around 500 B.C., the Greek philosopher Pythagoras, whose name has been immortalized in the field of mathematics, speculated that human life originated from a blend of male and female fluids, or semens, which came from within the body.

In the third century B.C. Aristotle, another brilliant Greek philosopher, suggested that these semens were purified elements of blood, and that the male semen formed in a woman's uterus by the coagulation of menstrual blood. Blood was thus thought of as the tool of heredity, and Aristotle considered that the female provided the "matter" and the male the "motion." The blood theory gave rise to phrases which are used in the Western world today, including such expressions as blood relative, bad blood, bloodstock and royal blood.

The Institutes of Manu, produced in India between A.D. 100 and 300, postulated the role of the female to be similar to that of a field, with the male role that of the seed. The joining of the field and the seed could give rise to new bodies. It is now known that both parents give their offspring the same amount of genetic information, and although some characteristics of one parent may dominate over those of the other, children bear a resemblance to both their parents, and thus to some extent to their grandparents, and so on.

The blood heredity concept persisted for another two thousand years until the seventeenth century when, in around 1651, the great English physician William Harvey finally disproved it. Harvey

The amount of corn harvested per acre in Europe in the sixteenth century (left) *was about one-tenth of that produced today. This is partly because of the previous lack of sophisticated machinery and partly because scientists now understand heredity sufficiently to breed species with a higher yield. It has been suggested, though, that as far back as Assyrian times, in the mid-800s* B.C., *farmers carried out cross-fertilization between different plant species, as depicted by some of the carved reliefs of that period* (below). *Alternatively, these scenes may depict instead a ritual in which date palms played some symbolic part.*

achieved fame for his discovery of the blood's circulation around the body, but he also showed that in deer killed at various times after mating there was never any sign of coagulation of menstrual blood. Instead, a tiny ball, or embryo, developed which gradually increased in size and complexity to resemble a deer. He concluded that the origin of the tiny ball was a small egg.

It was not until the latter half of the seventeenth century that the Dutch scientist Regnier de Graaf, who made significant contributions to the science of endocrinology, hit upon the idea that the union of egg and sperm is responsible for the miracle of producing a new being. He noticed small bulges, now known after de Graaf as Graafian follicles, on the ovaries of mammals. The bulges contained the unfertilized egg, or ovum, and inspired the idea that sperm was not the only agent responsible for carrying hereditary material. This new concept provided an answer to the puzzle of how female characteristics were transmitted to offspring, although it was many years before this was generally accepted by the scientific community.

Eighteenth-century Genetics

A little later, the naturalist Pierre Louis Moreau de Maupertuis, who was born in France in 1698 and who studied heredity in animals and humans, became convinced that both parents contributed equally to the characteristics of their offspring. He backed up his hypothesis with experimental proof. His theory that there were hereditary particles, that each parent provided particles which paired up to form a particular body part, and that all the pairings together formed the whole body was more than a century ahead of his time. Maupertuis suggested that, in any pair of particles, one could dominate the other, which explained how a child might resemble one parent more than another.

Maupertuis based his conclusions on detailed

The Graafian follicle contains an unfertilized egg and is located in a woman's ovary. Its discovery was the first indication that women as well as men contribute to the heredity of their offspring.

The giraffe's long neck and legs were erroneously thought by Lamarck to be inherited from an antelopelike animal which, in browsing on trees, stretched its body, and through only a few generations, became a giraffe.

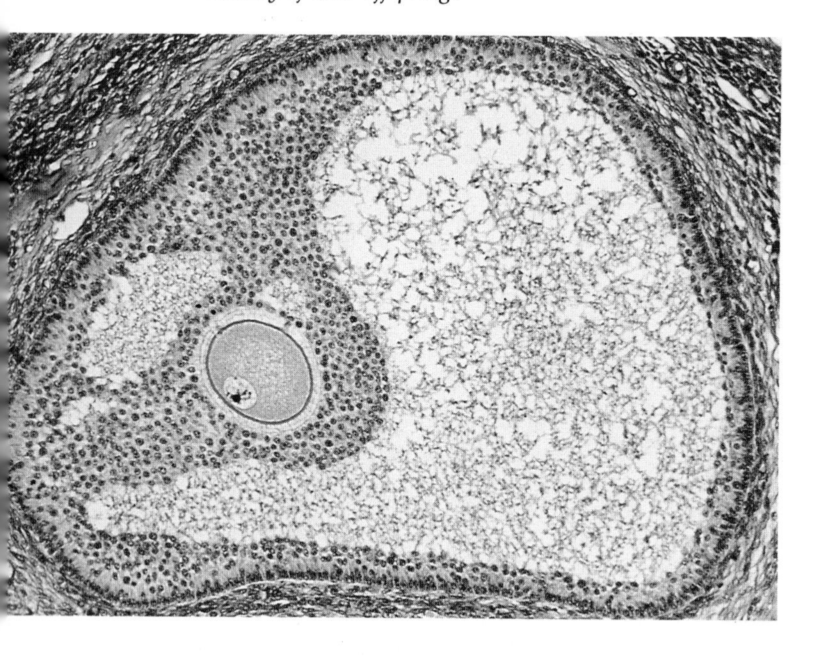

observation of the transmission of unusual characteristics, such as of a six-digit foot, through families. The inheritance of a six-digit hand he investigated in a German family. A German surgeon had inherited the characteristic from his mother; his father had exhibited nothing unusual in his hands or feet.

The surgeon then married a woman with no unusual features and they had six children, two of whom had the six-digit trait. Maupertuis thus supposed that six-digitism was transmitted by both father and grandmother. If he were alive today Maupertuis would be gratified to see how correct his ideas were, in that the dominance of some paternal genes over maternal ones and vice versa is closely analagous to his theory of hereditary particles.

While Maupertuis was making discoveries about human heredity, pioneering work was also being carried out on plant heredity, with important economic implications for farmers. Two famous botanists, the Swede Carl Linnaeus in 1760 and the German J. G. Kölreuter, carried out experiments in which they crossed different varieties and species of plants. They discovered that most hybrids were an intermediate of the parents, although some characteristics more closely resembled those of one of the two parents. From his experiments Kölreuter was able to conclude, just as Maupertuis had done in his animal experiments, that hereditary elements in plant progeny were derived equally from the parents. Aristotle's theory that one parent — the male — was responsible for the passing on of characteristics was thus further disproved.

Lamarckism

A new and different view of heredity came from the French naturalist Jean Baptiste de Lamarck (1744–1829). In 1809 his major work, *Philosophie Zoologique,* was published and contained some fascinating views on the inheritance of characteristics. Lamarck proposed that the constantly changing environment put organisms under stress. To survive, organisms had to change physically in an attempt to adapt to the environmental stresses. New structures would appear in response to changing conditions (such as the need to avoid a new type of predator or the need to search out new sources of food), whereas unused structures gradually disappeared.

Although he was a pioneer in observing the process of evolution, Lamarck's theory contained one point of contention: he believed that the adaptations to the environment were transmitted to the next generation. Thus if giraffes had to stretch their necks in order to reach leaves of trees during a drought, their offspring would be born with longer necks. This concept of the inheritance of acquired characteristics is known as Lamarckism.

At the time the theory seemed quite logical, but soon afterward other biologists detected a flaw in Lamarck's argument. Although they could not propose alternative theories, they discounted the idea that new forms of life could develop simply because of a response to environmental change. If Lamarckism were true, a parent who lost a limb could transmit that characteristic to his or her children, and similarly offspring could inherit large muscles from parents who had spent years in strenuous occupations.

Lamarck's ideas were effectively drawing a parallel between biological and legal inheritance, in which property that had been built up and improved could be inherited by offspring in the same way as human characteristics. In fact many languages employ the same words to describe legal

The Swedish botanist Linnaeus cross-bred plant species and found that most hybrids bore characteristics of both parents, although the contribution of one parent was often more obvious in the offspring.

and biological inheritance (although the two processes are evidently very different).

In 1580 the French writer Montaigne had questioned how he inherited gallstones from his father who did not have gallstones when Montaigne was born. And how was it that Montaigne did not inherit gallstones from his father until he was forty-five? In the same way as Lamarck's theory has been disproved, so Montaigne's question can be answered. Nobody inherits gallstones from the parents — but what a son can inherit is a set of genes, some of which may interact with the environment to predispose him to gallstones.

Charles Darwin

It was the English biologist Charles Darwin who in the second half of the nineteenth century attempted to lay Lamarck's theory to rest once and for all. Following what was arguably the most famous voyage ever taken by a biologist, Darwin's observations were published in 1859; the title of his subsequent book is now generally abbreviated to *The Origin of Species*.

Charles Darwin has perhaps had a greater influence on human thinking than any other scientist. He started (but never completed) a medical course at Edinburgh University and, after three years of study with a view to joining the clergy, he sailed on H.M.S. *Beagle* as an unpaid volunteer naturalist. During the five-year voyage to South America, Australia, and most significantly the Galapagos Islands, Darwin made many remarkable zoological and botanical observations. In the Galapagos Islands, a cluster of fifteen islands straddling the equator off the west coast of South America, Darwin noticed that over a long period of time populations of birds of the same species appeared to have migrated to other islands and evolved distinct physical differences which helped them adapt to their own island environment.

On returning home, Darwin deliberated for several years on what he had seen until, prompted by a manuscript from Alfred Russel Wallace, who had come to similar conclusions from his studies in the Malaysian islands, he published his views on the significance of the differences between the populations of the Galapagos Islands. Darwin's revolutionary conclusion was that each population

Darwin (right) *traveled on H.M.S.* Beagle (below) *to various parts of the world. The variety of species he saw inspired his theory of the survival of species by the natural selection of inherited characteristics.*

contains within it a number of individuals with a variety of physical differences, just as in a human population some people are tall, some short, some fat, and so on. Such differences meant that some individuals were better adapted to their environment than those who did not share the same characteristics.

Because there is a struggle for survival within all natural populations, only the fittest survive. The survivors thus dominate the reproduction of the population, and transmit the favorable characteristics to their offspring. Those who are less fit die, their "less fit" characteristics being lost to the population. Darwin called this survival of the fittest "natural selection." Beneficial traits thus become the norm in a population.

Of course this perception did not take place overnight; it took Darwin a great deal of diligent observation, followed by further exhaustive studies on inbreeding, for him to arrive at his conclusions. Darwin's reputation had already been established by his first book published in 1840, entitled *The Zoology of the Voyage of the Beagle*, and when the first edition of *The Origin of Species* was published it was sold out in a single day.

Darwin's work raised a furor throughout the

The lawyer Clarence Darrow (in shirt sleeves) defended Professor John Scopes in the celebrated trial in which he was convicted for teaching Darwinism. Darwin's theory of evolution remains contentious.

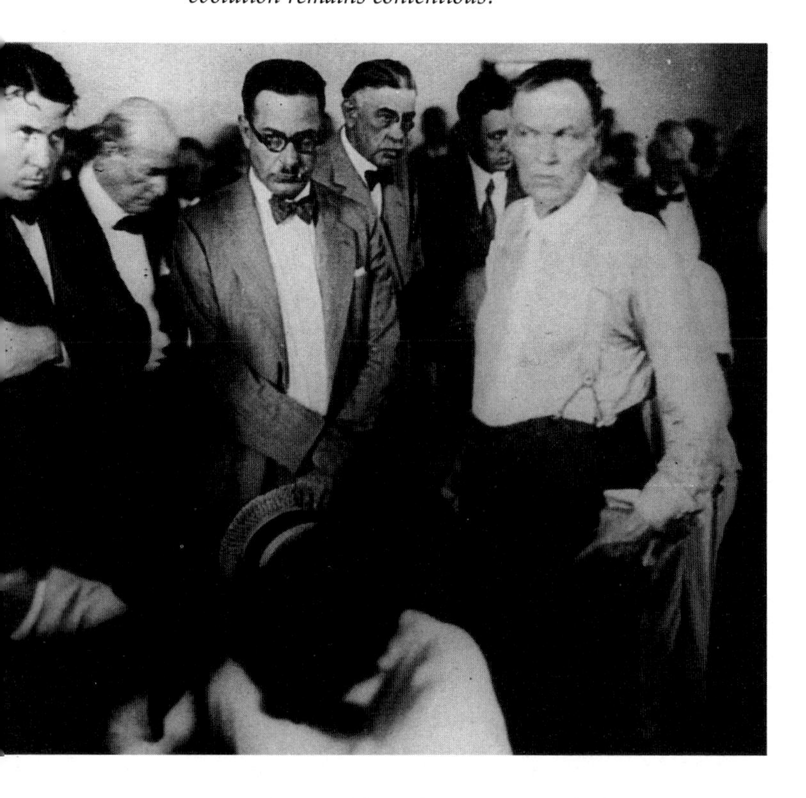

scientific world and was particularly despised in some religious quarters because Darwin asserted that evolutionary change occurred not by design but by a random gradual progressive process, dependent on chance variability within populations and influenced by the environment. In his book *The Descent of Man*, Darwin extended his theories to propose that the existence of all species, including the human race, arose by chance, not by design. This idea could be extrapolated to mean that humans had evolved from "lesser" animals such as apes, to whom humans bear closer resemblance than to any other species on earth.

It is easy to understand how, in view of this last point, Darwin's theories were rejected by many who refused to accept the idea that the forefathers of man had been apes. In addition there was an obvious clash with the religious view that Adam and Eve were the first on the planet and all humans were direct descendants from them. The whole pattern of religious belief had thus been challenged, and even today there are many who accept only the biblical version of Creation.

Resistance to Darwin's ideas continued for many years after his death in 1882, into the twentieth century, even though much irrefutable evidence supporting the theory of natural selection had by then been accumulated by other scientists. One famous example of resistance to Darwinism occurred in the 1925 "Monkey Trial," in which biology teacher John Scopes was convicted for teaching Darwinism in his Tennessee classroom, breaking a Tennessee law which forbade inclusion of the theory of natural selection in biology classes. The courtroom battle between the opposing lawyers, William Jennings Bryan for the State and Clarence Darrow for the defense, has been recreated in literature, in the theater and in films.

Gregor Mendel

While Darwin was looking at inheritance in an evolutionary context the remarkable Moravian monk Gregor Mendel was studying how offspring inherited specific sets of characteristics from their parents. Some scientists describe the work of Mendel as the starting point for modern genetics, although it was many years before the significance of his work was realized. When Darwin died more than fifteen years after Mendel's astounding revelations had appeared in a scientific journal, he had never even heard of Mendel.

Mendel, having failed in his attempt to become a teacher, decided to devote himself to the study of plants. He obtained permission to cultivate some plants in the garden of the monastery in which he lived — a bold, and possibly secret, move for the local bishop would not even allow monks to teach biology. But, having been stimulated by the biologist Franz Unger while at the University of Vienna, it was Mendel's aim nonetheless to unravel some of the secrets of plant heredity. Accordingly, he is thought to have started his famous experiments on peas in around 1856. His intention in experimenting was, on his own admission, first to try to predict what different forms of hybrids would appear in different generations and to analyze by statistics their frequency of occurrence.

Mendel decided to use the garden pea for his experiments and chose seven pairs of contrasting characteristics for comparison, such as round or wrinkled seeds, tall or dwarf plants, the color of the plant's flowers, and so on.

Gregor Mendel

Hybrids and Heredity

In an all-time list of geniuses forgotten by their own age, the Abbot Gregor Mendel would surely come at or near the top. What he achieved in his experiments in the garden of the abbey where he was a monk was far ahead of his time. Had it been appreciated during his lifetime, the science of genetics might have been at least forty years older now. But it was not. Not only did his work fail to attract any attention at the time, but at his death his papers were burned. Fortunately, enough of his work was published for his studies to be acclaimed at their true worth after his death.

Mendel was born into a rather poor farming family at Heinzendorff in Moravia (then Austria, now Hynčice in Czechoslovakia) in July 1822. Despite this background he managed to gain entrance to Olmütz University. After two years, however, his money ran out, and he joined the Augustinian order at the monastery at Brünn (now Brno). He was then aged 21. Four years later he was ordained as a monk.

The abbot was sympathetic to Mendel's academic yearnings and encouraged him to try for a teaching diploma at the University of Vienna (he was already teaching at the local school). But, being a nervous student, he was poor at examinations, and failed, returning to the monastery (which was a teaching order). On his return from Vienna in 1854 Mendel decided to devote his monastic seclusion to plants, having been fascinated by them during his youth on the family farm where his father crossed and grafted crops to improve them.

He took to hybridizing plants, especially peas, by such thorough methods that he was able to use the results for a considerably deeper insight into the workings of heredity than anyone had yet achieved.

Thus Mendel derived his two great laws. The first stated that where two plants of the same species were crossed, each with one different trait, the first generation of offspring would all be the same in appearance—all displaying the "dominant" trait of the two originals. But in the second generation three plants would appear displaying the dominant trait for every one that appeared with the "recessive" trait; there would be a ratio of 3:1.

Mendel's second law—of independent assortment—was similar. When two plants of the same species, but having two different traits, were crossed, the first generation were again all the same (displaying the dominant trait)—but the second generation resulted in nine of the dominant type, three each of two intermediate forms, and again one of the recessive type; the ratio was 9:3:3:1.

His innovatory discoveries received with indifference by his contemporaries and his eyesight failing, Mendel may have been glad to succeed as abbot in 1868, and to immerse himself in administrative duties.

He died in January 1884. Almost twenty years later his work was rediscovered and he was at last accorded the status he deserves.

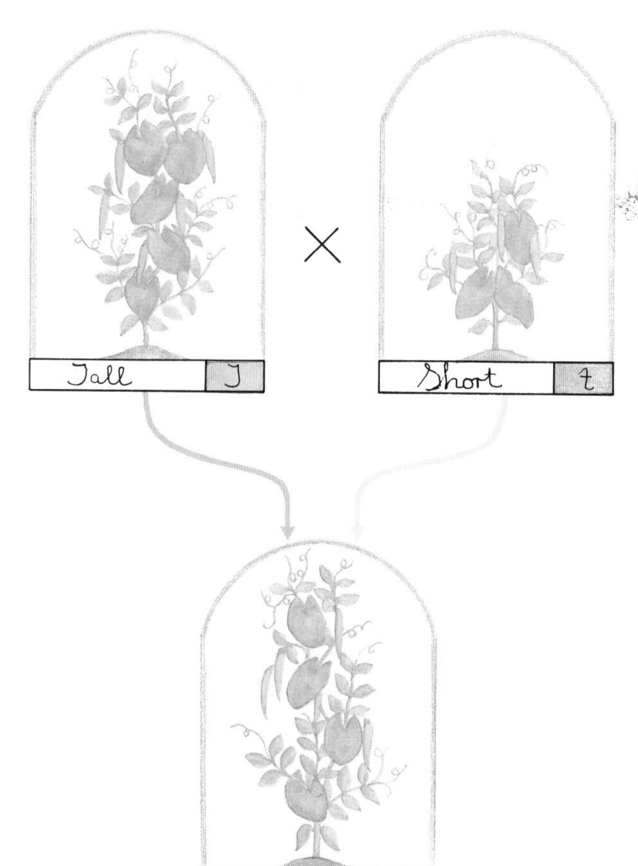

To understand fully what Mendel did it is necessary to set out in detail one of his experiments carried out in the monastery garden: hybridization between a tall and short plant, for example. If the traditional view was to be believed — that hybrid characters fall somewhere between those of their parents — then the new plants should have been of medium height.

Mendel radically opposed the traditional stance; following literally hundreds of initial experiments he had his own theory. He maintained that each simple pea characteristic was under the influence of two particles (which we now call genes), that each parent donated one particle to the offspring. And Mendel theorized that if the two particles were different, one (the dominant) would dominate the other (the recessive) in terms of expression of characteristics. Crossing a tall and a short plant was Mendel's way of testing his hypothesis. If he was right, the first generation of hybrids would all be of uniform size, tall or short.

Firstly, to prevent the short plant fertilizing itself Mendel "doctored" or emasculated it. This enabled him to carry out "artificial insemination" by pollinating it with pollen from the tall plant. The pollen contained pollen nuclei, the equivalent of sperm in an animal, which fertilized the ovules (eggs) of the short plant. Subsequently, pods appeared on the plant. Their seeds were then planted — and, perhaps to his own amazement, Mendel's theory proved completely correct: all the plants grew to be tall. In today's genetic language we would thus say that tall is dominant over short.

The second stage of Mendel's brilliantly innovative work was to note the results of producing a second generation by fertilizing the hybrids with their own pollen, allowing the pods to grow and planting the seeds. Now according to Mendel's theory, the first generation hybrids would be made up of particles (or genes) which contained combinations of the dominant and recessive particles. Again after many initial experiments, he evolved an extension to his first theory, to the effect that by recombining these particles, some of the offspring were dominant plus dominant, others were dominant plus recessive, and there were also those which were recessive plus recessive. Simple mathematics shows that the second generation should therefore produce three tall plants to every short plant. This is in fact the famous Mendel ratio of one out of four (or one to three) that has immortalized his name in genetics.

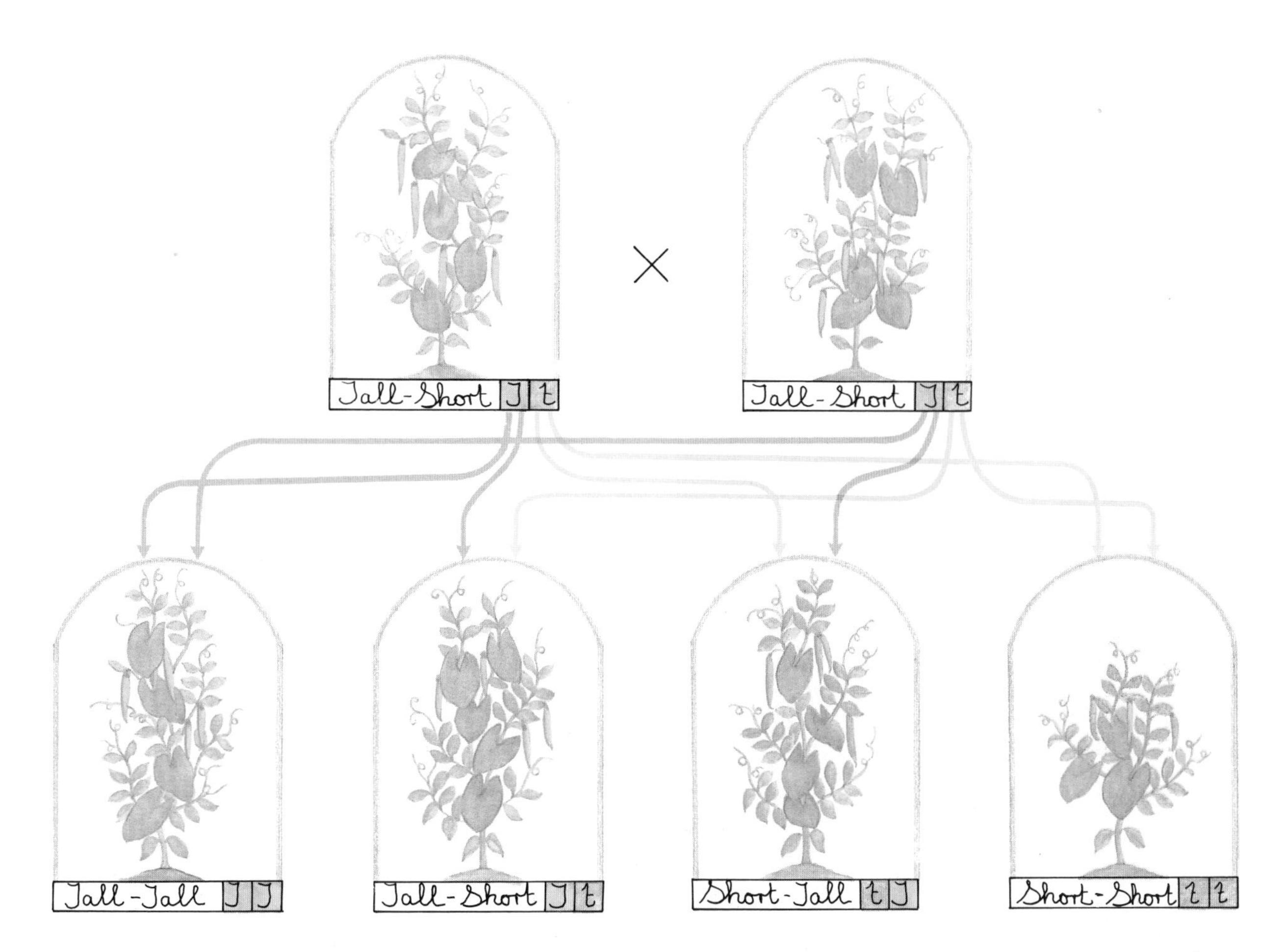

Now redesigned, the garden at Brno monastery in Czechoslovakia (far left) *was the place where Gregor Mendel performed his famous experiments on inheritance, using garden peas. Crossing a tall plant with a short one* (left) *he bred a plant that was tall, having the gene for tallness (which is dominant) but also containing the gene for shortness (which is recessive). Known as F_1, this plant was crossed with another F_1 hybrid* (above). *The appearance of the resultant F_2 offspring was tall to short in a ratio of three to one, although their genetic constitution was tall-tall, tall-short, short-tall and short-short.*

Mendel's further results proved his extended theory. He presented his results to the Natural History Society of Brünn in 1865 and the following year they were published in the Proceedings of the society. But he might as well not have bothered, for at the time nobody cared or understood what the monk was talking about. If he had been a professional scientist he might possibly have been able to get his results published in a better-known scientific journal, to be read by botanists and biologists in other European countries. He tried sending reprints of his paper to scientists abroad, but they were largely ignored.

Two years after the publication of his results something happened to halt Mendel's garden experiments — he was elected abbot of the monastery — and from then onward he spent the rest of his life expending considerable energy in its administration and in trying to persuade the government that monasteries should not be taxed. He also turned his attention from breeding plants to breeding bees, and produced a new strain of bees which gave magnificent honey. (Unfortunately, they had all to be destroyed because they stung everybody for miles around.) Mendel in addition developed an interest in meteorology.

The two types of primrose at the top, Primula veris *and* P. vulgaris, *were crossed to produce the flower in the middle, from which the bottom plant, the garden flower known as polyanthus, was derived.*

Although Mendel's colleagues at the monastery did not like the experiments he was carrying out or his interest in the new biology, which included Darwinism, the Czech revolutionaries he sheltered in his monastery became very fond of him. When he died in 1884, the famous Czech composer Leoš Janáček played the organ at his funeral.

Soon afterward the monks elected a new abbot, who, tragically for succeeding generations, burned all of Mendel's papers. All that is left of his work is the paper he produced for the local natural history society. Mendel's scientific achievements were like those of many great artists: the value of his work was appreciated only long after he had died.

In Mendel's case his work remained unknown for nearly fifty years because his contemporaries were preoccupied with Darwin's theory of evolution. His laws of heredity remained in obscurity until 1900 when remarkably, within the space of just a few months, they were rediscovered independently by three botanists: Hugo de Vries, Professor of Botany at the University of Amsterdam, Carl Correns, a botanist at the University of Tubingen, and Erich von Tschermak-Seysenegg, an assistant in the agricultural experimental station at Esslingen near Vienna. It is thought that all three of them were inspired by reading Mendel's work. From that time Mendel became one of the great names of biology, so much so that the laws governing the general inheritance of characteristics are commonly referred to as Mendelian.

Identical twins have been invaluable to the study of the effects of heredity and environment on people. It is now known that environmental factors play an important part in shaping personalities.

After Mendel

While Mendel was investigating plant heredity, Sir Francis Galton, a versatile genius and cousin of Charles Darwin, became interested in human inheritance. Like Darwin, Galton began his career as a medical student but departed from medicine after being left a substantial legacy. Just as Darwin showed that the environment played a role in animal development, so Galton used human twins to show the difference between development as a result of heredity and development as a function of the environment. He argued that identical twins must have the same genetic constitution, and that any differences which appeared during their development resulted from the effects of the environment. Galton's conclusions stemmed from a series of questionnaires he sent to the families of twins, asking for their life histories. His findings were first published in *Frazer's Magazine* and the *Anthropological Journal* in 1875, and later in his book *Inquiries into Human Faculty and its Development*.

Galton also put forward the controversial theory, in which some still believe today, that intelligence is entirely inherited and is not solely a function of the environment, the effect of cultural and wealth advantages. Galton advanced the idea of hereditary improvement in men and animals by the use of selective breeding — that is, the encouragement of mating between what he regarded as the best of human stock — and he coined the word "eugenics" to describe such a process. Eugenics in plant and animal breeding has led to the development of new crops and species which have helped man to farm and to produce food more efficiently. The application of eugenics to humans is, however, deeply controversial. There are numerous ethical

problems associated with its practice — such as deciding which are desirable traits and which are not, and who would make such a choice.

A contemporary of both Mendel and Galton was the German biologist and philosopher Ernst Haeckel, who looked at the development of embryos. He put forward the theory that heredity was just a simple continuation of growth, and that cells contained particles endowed with memory passed on from one generation to the next, where they expressed their activity. Haeckel may have come to this conclusion from the observation of unicellular organisms which multiply by dividing in two (fission), and in this sense are "growing."

Haeckel also believed in the theory that "ontogeny recapitulates phylogeny" which, simply stated, means that the early development of embryos seems to repeat the adult stages of lower forms of life from which it has evolved.

Haeckel's ideas can no longer be accepted in a simplistic fashion, but at the time attracted great attention. His theory differed from that of Darwin, whose mistaken beliefs on reproduction are often ignored. Darwin's explanation of fertilization, at a time when the mechanism by which sperm fertilized an egg was not well understood, was derived from the ancient Greek idea of pangenesis. According to this theory, every organ and tissue secreted granules called gemmules, which combined to make up the reproductive cells, almost as if they were ambassadors being sent to represent, and in some way constitute, the cell in following generations.

It was the German physician August Weismann who, late in the nineteenth century, showed that the reproductive cells (germ plasma) developed independently of other body cells, and that bodily changes do not affect them. (This was yet further evidence to discount Lamarck's notion of the inheritance of acquired characteristics.)

The German biologist and philosopher Ernst Haeckel (below right) *was studying cell division, or fission (now known as mitosis,* below left) *at the same time as Gregor Mendel and Francis Galton were considering the question of heredity. He suggested that during cell division the "memory" of the parent cell passed into the daughter cells, so continuing heredity through the process of growth.*

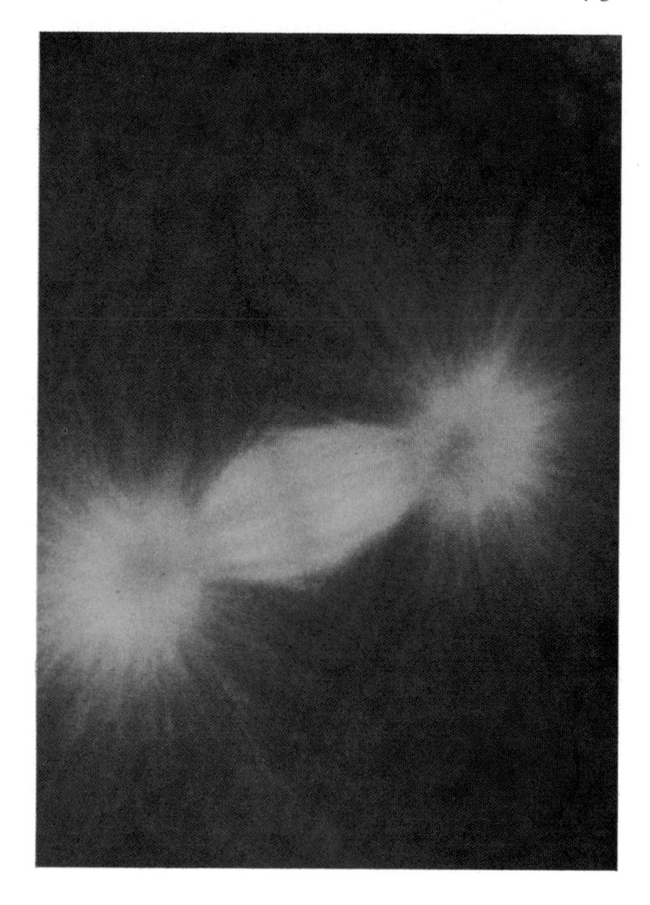

Weismann was one of a number of great scientists during the latter half of the nineteenth century who were fascinated by the process of heredity. After the year 1900 progress in genetics accelerated. The man credited with first coining the term "genetics" was William Bateson.

Bateson was an English zoologist who carried out a number of plant breeding experiments and he came to nearly the same conclusions as Mendel. Perhaps Mendel — his mind uncluttered by the wide spectrum of scientific teachings — was better able to understand how inheritance might work, and thus to extrapolate the Mendelian theories. Yet fate took a turn in Bateson's favor. While he was traveling on a train to London (to present a paper on his experiments to the Royal Horticultural Society) he read Mendel's paper for the first time, and from that point he became a devoted disciple. He translated Mendel's paper from German into English and it was published in the *Journal of the Royal Horticultural Society* in 1900; this translation helped put Mendel's views in front of the whole scientific world.

Bateson saw how Mendel's theories explained many of the results of his hybridization experiments and those of others. He himself carried on for many years with experiments in breeding, not only looking at traits in plants but also in chickens, and his enormous contributions to classical genetics finally gained its acceptance, along with much of Mendel's ideology, as a recognized discipline.

Genetics Advances into the Twentieth Century

Shortly after Bateson's "awakening" to the work of Mendel one of the major advances in genetics took place. At that time, scientists knew that each animal or plant cell contained a nucleus and that within the nucleus were a number of minute threadlike structures called chromosomes. But it was not until 1903 that two scientists, the American Walter S.

The breeding of animals such as chickens for greater yields and meat of improved quality has interested many scientists, including William Bateson who experimented on breeding traits in the early 1900s.

Sutton and the German Theodor Boveri, independently suggested that the chromosomes were the carriers of the information which offspring inherited from their parents. They said that chromosomes exist in homologous pairs, one from the father and one from the mother in each pair. And they went on to declare that chromosomes carried the hereditary factors, or genes, and that the behavior of chromosomes at cell division explained how characteristics were inherited.

Although men like Sutton and Boveri applied their thinking to human inheritance, most genetic work had by that time been carried out on plants. This situation changed dramatically when W.E. Castle, a pioneer of genetics in the United States, introduced to the laboratory an animal that was to be one of the major tools in the study of genetics for many years to come. The animal was the fruit fly *Drosophila* which has a number of features that make it an ideal subject for studying genetics. Even today universities use the fruit fly for research and to teach biology students in practical experiments.

Fruit flies make such good genetic models because they are easily bred in the laboratory; their life cycle takes about nineteen days, which means that many generations can be studied in a year (in humans the same number of generations would take more than 750 years); they are cheap to breed; and the female produces thousands of eggs during her short lifetime. *Drosophila melanogaster*, the species most commonly used for experimentation, also has the advantage of having four pairs of chromosomes which are easily identifiable under the microscope. An added advantage is that the chromosomes in the salivary glands of the fruit fly larvae are among the largest found in nature, being more than one hundred times larger than the chromosomes in most other cell types. The reason for this is not clear — but it allowed geneticists to see distinctive patterns of transverse bands across

Philip IV, King of Spain, painted astride his horse by Rubens, was one of the far-flung Hapsburg clan, Spanish members of which inherited their own most noticeable trait, the long, jutting chin.

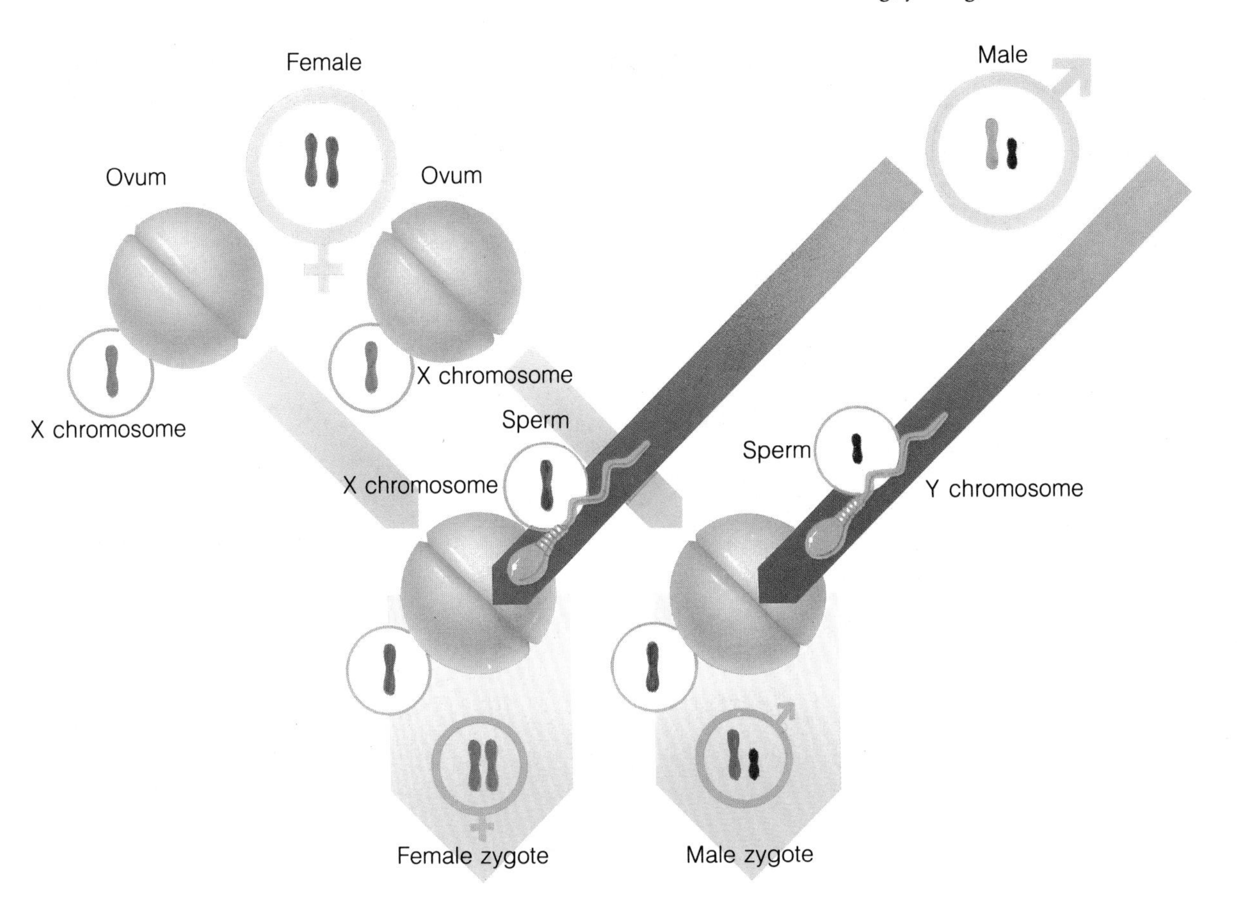

The sex of a baby is determined at fertilization by the chromosomes inherited from the parents. In meiosis, the division of sex cells, a female cell (which contains two X chromosomes) separates to give sex cells that have one X chromosome each. Similarly a male cell (which holds one X and one Y chromosome) divides so that some sperm contain one X and others one Y chromosome. When a sperm with an X chromosome fertilizes an egg cell a female cell is formed again, which becomes a female individual. The combination of a female X and a male Y chromosome forms a male cell, which develops into a male individual.

the chromosome. It has since become possible to localize individual genes to a small group of bands or in some cases a single band.

The American geneticist Thomas Hunt Morgan and his students Calvin Bridges and A. H. Sturtevant were among the first to study the genetics of fruit flies. They established the fact that hereditary units, or genes, are arranged in a particular sequence along the chromosome. Using subtle breeding techniques, Morgan was able to discern not only the order of genes on a chromosome, but also the distance between them. Morgan continued his fascinating work on breeding for many years and was awarded a Nobel Prize in 1933, the hundredth anniversary of Alfred Nobel's birth. Totally in character, Morgan excused himself from the magnificent centennial banquets and waited until the following year before going to Stockholm to give his acceptance speech.

It was around the same time, in 1909, that the

A fruit fly feeding on a rotting apple (below left) *may appear to be an insignificant bug, but these insects have been of tremendous importance in the study of heredity and mutations. One reason is that the salivary glands of their larvae contain some of the biggest chromosomes found in any animal. These chromosomes* (below right) *have enabled closer investigation of banding and gene locations.*

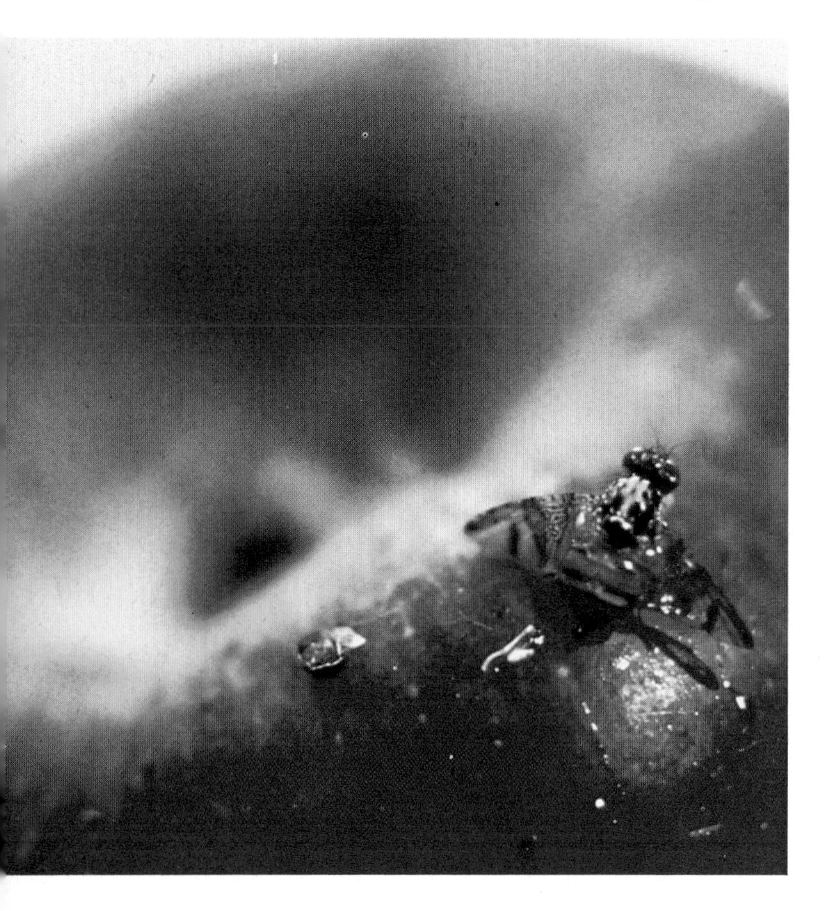

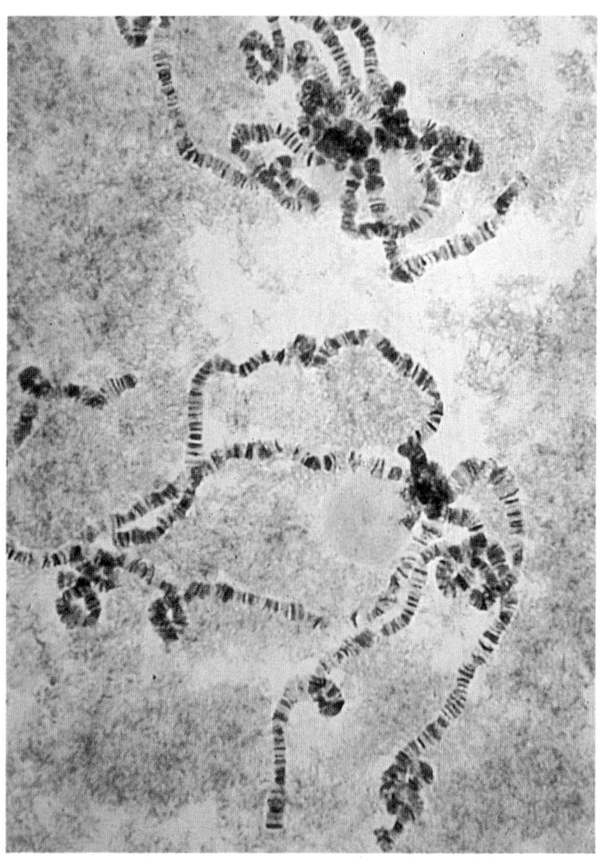

Danish botanist Wilhelm Johannsen identified two important genetic concepts. He was the first to make clear the difference between genotype, the genetic constitution, and its physical expression, the phenotype, which is also affected by the environment. The phenotype changes continuously (as the environment changes) during life, but the genotype remains the same, although the genetic constitution of an adult's cells are copies of the original chromosomes present in the fertilized egg. Johannsen was also responsible for coining the word "gene" for hereditary factors. The genes responsible for contrasting traits are known as allelomorphs, or alleles for short.

While Johannsen was busy on his genetic investigations — which had been inspired by Galton — an English mathematician, G. H. Hardy, and a German physician, W. Weinberg, formulated one of the most important principles of genetics: the Hardy-Weinberg equation. They applied mathematical theory to genetics to predict how genes would spread within a population, and hence determine the distribution of dominant and recessive characters. It might be expected that dominant traits would eventually take over at the expense of recessive ones, because by Mendel's laws there will be three dominant progeny to one recessive. But the Hardy-Weinberg principle shows clearly that in a large, randomly-mating population undisturbed by outside influences such as predators, dominants do not increase at the expense of recessives. The proportions of dominant to recessive genotypes remain constant from one generation to another (although mutation can change this principle among small populations).

One of the next major advances in genetics came in 1927 when Hermann J. Muller, a former student of Thomas Morgan, showed that genes are able to change if they are exposed to X rays. He called this "mutation," and increased the mutation rate of

Thomas Hunt Morgan

Mendel's Reluctant Heir

The importance to modern genetics of the work of Thomas Hunt Morgan would be difficult to exaggerate. Through his research and experimentation—in which he studied and popularized the use of the fruit fly *Drosophila melanogaster* as an experimental animal—he proved the theory that chromosomes were responsible for hereditary effects, that combinations of genes were responsible for individual characteristics, and that such characteristics could be traced to specific gene combinations which could then be charted on a new type of map. Away from the field of genetics, his first and last research was, however, into embryology.

Morgan was born to a prominent family in Lexington, Kentucky, in September 1866. At the age of twenty he graduated from the State College of Kentucky and four years later received his doctorate from Johns Hopkins University. In the following year, 1891, he became Associate Professor of Zoology at Bryn Mawr, from where, three years afterward, he transferred to Columbia University as Professor of Experimental Zoology. Finally, in 1928, he was appointed Director of the Biological Sciences Laboratory at the

California Institute of Technology (Caltech).

It was during his time at Columbia University that Morgan formed his first conclusions about the newly rediscovered Mendelian laws of heredity. He thought little of them. Chromosomes had been related to Mendel's "factors," but it was evident to all that there were far more hereditary variations than there were chromosomes to account for them. And if one chromosome was responsible for a number of traits, all those traits must surely appear without variation in succeeding generations, because chromosomes were directly transferred from parent to child.

For Morgan, the crowning piece of evidence came after long experimentation with the fruit fly *Drosophila*. He found—in his attempts to disprove Mendel—that "white-eye" mutant fruit flies were always male. A sex-linked mutation was thus possible—indicating a much greater significance than had previously been accorded to the difference between the chromosomes of males and females. Further investigation confirmed the presence and importance of genes, and also suggested that the genes were arranged in linear fashion on the chromosome. Mendel had pretty well been right—but his "factors" were genes, not chromosomes.

From these data, Morgan's colleague A. H. Sturtevant compiled the first chromosomal map, analyzing groups of linearly-linked genes on individual chromosomes.

Thereafter, although work in genetic research continued, Morgan concerned himself more with the experimental embryology that had been his first subject of study, assisted by his wife who was a cytologist.

In 1933 he received the Nobel Prize in medicine. He died twelve years later.

The Soviet biologist Trofim Lysenko believed, much like Lamarck, that an organism in abnormal environmental conditions develops to take advantage of them, and its offspring inherit this capacity.

In the early 1900s it became clear to scientists that an individual's appearance is formed from inherited traits as well as the environment. The girls to the left are about the same age as the one above, but their harrassed, lined faces give them the appearance of being older.

fruit flies fifteen thousandfold over the natural spontaneous mutation rate by using X rays. (Muller received a Nobel Prize in 1946 for his work in the field of radiation.) Muller believed, correctly as it turned out, that an X-ray beam causes some sort of damage to the genetic material, and that this leads to a change in the cell's chemical reactions — which in turn leads to mutation.

In the 1930s Muller decided to work in the Soviet Union, where he met the remarkable geneticist T. D. Lysenko. It was Lysenko who made his special brand of Lamarckism (inheritance of acquired characteristics) the accepted view in the Soviet Union between 1938 and 1963, and caused the suppression there of teaching and research in modern genetics. Lysenko and his supporters published hundreds of books and articles to prove their contentions (which included the denial of any special properties to chromosomes and a refusal even to acknowledge the existence of genes), but they were discredited in 1964 — more than a hundred years after the publication of Darwin's *The Origin of Species*. Muller's experiments with radiation proved Lamarck — and by inference Lysenko — to be wrong. But Lysenko would have none of it, and the dispute drove Muller, a staunch supporter of eugenics, away from the Soviet Union.

The Investigation of the Gene

There still remained the question "How does a gene operate?" A series of breeding experiments by two American scientists, George Beadle and E. L. Tatum, showed that genes control the synthesis of enzymes, proteins which act to catalyze biological reactions within the organism.

The next piece of genetic detective work was to identify the substance in chromosomes which stores and thus transmits the genetic information. The answer came from two rather shy, meticulous men who identified DNA (deoxyribonucleic acid)

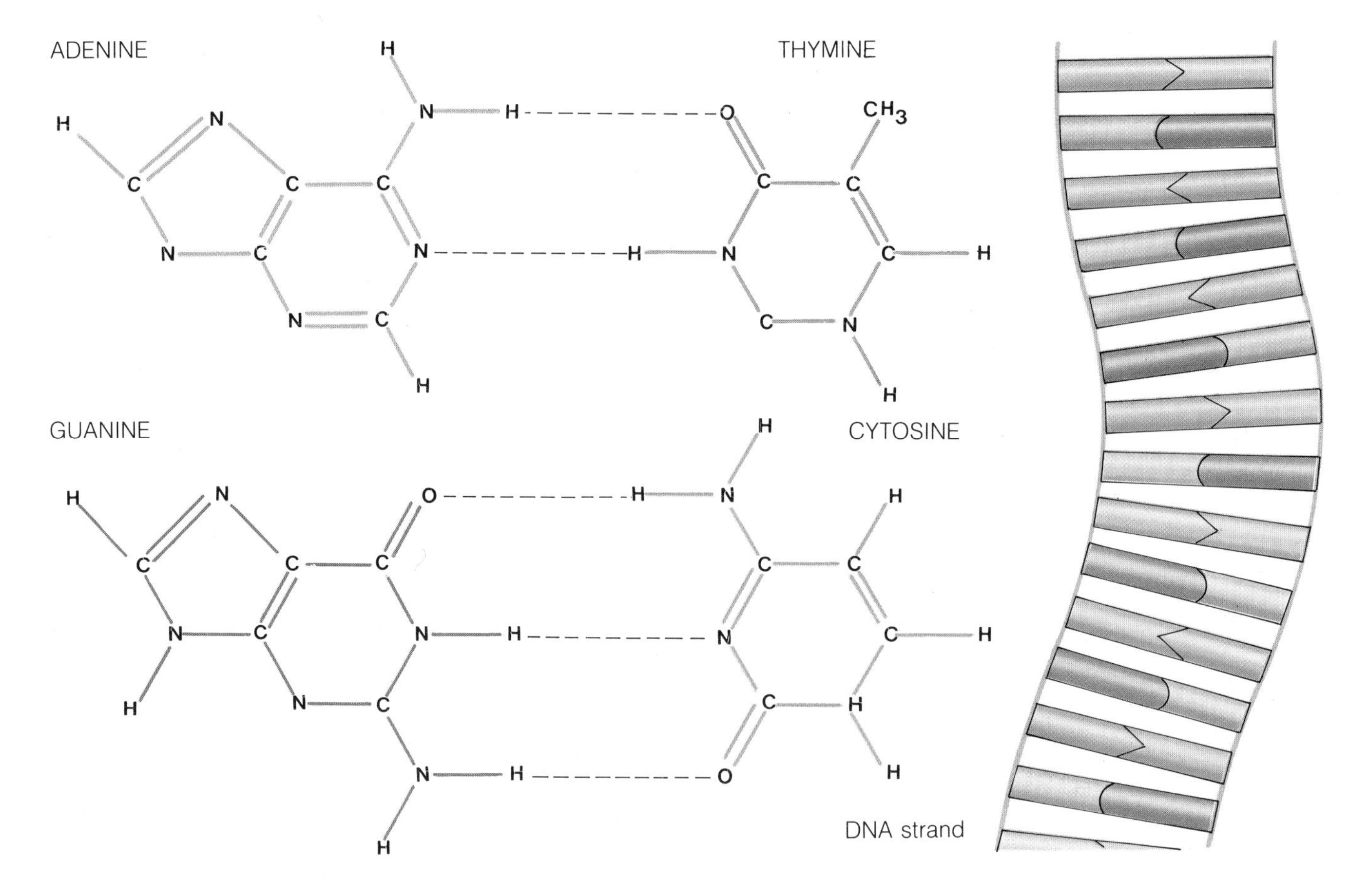

DNA was revealed in Watson and Crick's famous paper, published in 1953, to be formed from bonded pairs of bases: adenine and thymine, and cytosine and guanine. These bases pair themselves by hydrogen bonding to form the "rungs" of the spiraling ladder that is DNA, the storehouse of genetic information.

as the hereditary material. One of these men was Fred Griffith, a taxonomist at the Ministry of Health in London, who developed techniques to classify disease-causing organisms (pathogens). In 1928, Griffith performed a remarkable experiment using pneumonia-causing bacteria (pneumococci). These bacteria can produce two types of colonies on a growth medium — rough (R) and smooth (S). Griffith showed that if an S colony was killed using boiling water and the remnants mixed with R bacteria, then some of the R bacteria were converted to S bacteria using this non-living extract — and the change was a permanent, inheritable one. Thus a substance from the dead bacteria had been able to change the genetic nature of the R bacteria.

Griffith's experiments were largely ignored at the time, but they were repeated in the 1940s by Oswald Avery, the son of an English pastor. Avery left England for Nova Scotia and then settled in

The twisting, spiraling model of DNA shown here resembles a piece of modern sculpture whose form could be said to express continuity and the repetition of elements — the very nature of heredity.

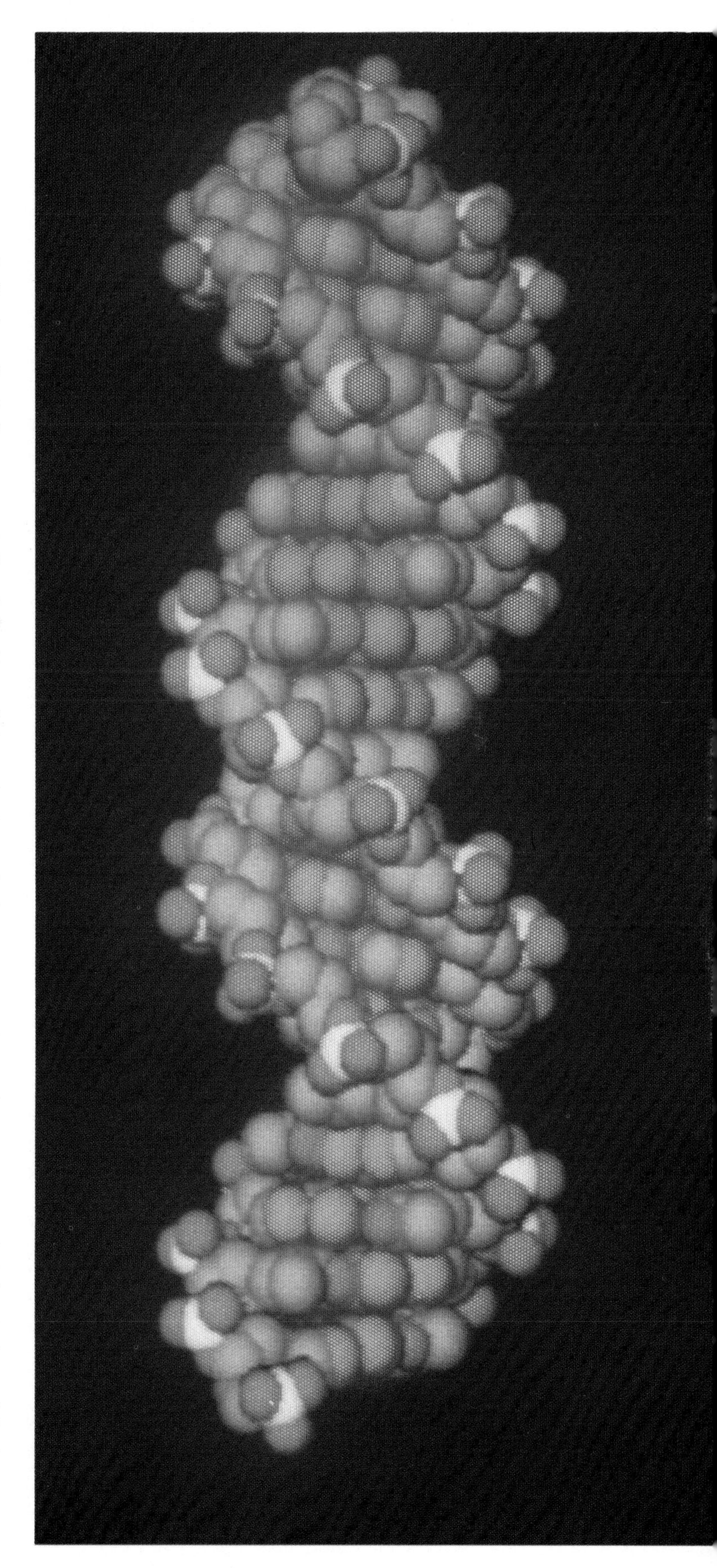

New York and worked at the Rockefeller Institute. He concluded that the transforming agent Griffith had talked about was a nucleic acid. He purified the remains of dead S bacteria and demonstrated transformation using only the nucleic acid fraction. In 1944 Avery, along with his research colleagues Colin MacLeod and Maclyn McCarty, published a now famous paper which described DNA as the genetic material.

The Investigation of DNA

The next great advance in genetics came with the discovery of the structure of DNA, a story full of drama. It started in 1951 when James Watson, a biochemist from Chicago, aged twenty-three, went to carry out research at the Cavendish Laboratory in Cambridge University in England. Watson became obsessed with discovering the structure of DNA and his enthusiasm infected Francis Crick, a little-known physicist.

Meanwhile at King's College, London University, a group of physicists led by Maurice Wilkins were also trying to unravel the secrets of DNA's structure. It was their work which provided much of the experimental evidence upon which Watson and Crick could test their theories. In the end it was Watson and Crick who published their paper in the journal *Nature* on April 25, 1953. In fewer than one thousand words they offered "conclusive evidence for Darwin's theory of heredity," and the pathway to modern molecular genetics was opened.

They wrote that DNA had a double-helix structure composed of a sugar-phosphate backbone held together by hydrogen-bonded base pairs. The structure was thus like a twisted ladder with the sugar-phosphate backbone as the sides and the pyrimidine-purine bases and hydrogen bonds the rungs. One type of base always pairs with a certain other type of base, the pairing always being equivalent.

The identification of this structure also provided an answer to how genetic information is passed from one cell to its offspring. The Watson-Crick model postulated that at nuclear division the ladder unzipped along its length and a complementary strand was made to match each of the two strands. Thus the two new double helices formed would be identical to the original double helix.

Chapter 2

Inherited Possessions

Like a hand dealt from a huge deck of cards, hereditary components are distributed randomly among their recipients. Progeny are each dealt half of the combined parental deck. Any hand may arise, but the number of arrangements is so vast — many trillions of genetic combinations are possible — that there is virtually no chance that two people will have the same genetic constitution. The only exceptions are identical twins.

Why are you yourself and not someone else? After all, you share genes with your relatives, and have half of each of your parent's genes. You are unique because the combination of your genes is unique, and nobody has exactly the same combination (even if you are an identical twin — although between such twins there are only subtle differences). Genes provide the fundamental information that determines how a person develops, although environment acts to shape the final product from the moment of conception onward.

Environmentally-influenced characteristics cannot be inherited. If the children of weightlifters could inherit an environmentally-shaped characteristic, they would all have larger muscles than everyone else — but this does not happen. A child may, however, inherit from the parent the potential to develop powerful muscles, and the environmental effect of intensive training could fully develop the inherited tendency.

The Story Unfolds

Gregor Mendel discovered the essential rule of heredity: inheritance is governed by the activity of discrete units and is not a simple blending of parental characteristics, like the mixing of paints. The units are now called genes, and to a large extent a person can be said to be a product of all the genes he or she carries.

It may seem strange that the full story of human inheritance has been unraveled from the study of seemingly insignificant animals or microorganisms — mice, fruit flies, bread molds and bacteria (particularly a bacillus called *Escherichia coli*, which lives in the human colon). But genetic experiments require a number of generations or breeding cycles to occur before the results can be determined.

The fruit fly (*Drosophila melanogaster*) has a short generation time of only nineteen days from egg to full maturity and the start of another generation, making it possible for a scientist to carry out a

Breeding experiments with animals such as the monkey (below) *can provide scientists with much useful information about heredity because of their genetic similarity with humans and because their generation spans are relatively short. Experiments such as the one in which a horse gave birth to a donkey foal and a donkey gave birth to a horse foal* (bottom) *may lead to a better understanding of the mechanisms of heredity.*

number of experiments on successive generations within a year. Mice have a generation time of nine to ten weeks, whereas monkeys — although they more closely resemble human beings than other animals used in genetic studies — have a generation time of more than a year. The information is potentially more interesting, but the process of obtaining it is not practical. And as far as human breeding experiments are concerned, even if moral, ethical and legal considerations could be ignored, the project would be defeated by the approximately twenty-year gap between generations.

Small organisms are also used because they can be studied in large numbers. Genetics is a study of random, sometimes rare, events. The cost of keeping an adequate number of larger animals would soon bankrupt most research laboratories.

Once a scientist has established genetic rules through experiments with animals, the rules can be applied to human beings by looking for examples of human inheritance that resemble the results of the experiments. Both "normal" characteristics and genetic disorders shed light on the patterns of human heredity and confirm that the laws of genetics probably apply to all creatures.

The simplest rule is that a person has two copies of each gene, and that one copy is inherited from the mother and one from the father. When genes are passed to offspring, either of the mother's two copies may be given to her child or children. The same is true of the father's contribution. The selection of copies that the child gets is a matter of random chance. Because of this, heredity must be studied on the basis of mathematical probability. For example, a genetic counselor advises parents on the possibility of having an abnormal child, using the mathematics of random chance.

Patterns of Inheritance

Mendel noticed that genes — which he called heredity factors — always came in pairs, but that not all gene pairs were equal. For example, one of a pair of genes might be dominant and its effect was always seen when either one or two copies were present. The other gene of the pair might be recessive, its effect masked if the dominant gene was present. This recessive gene would become apparent only when two occurred together.

Dominant inheritance occurs when one parent has one gene of a pair which is dominant. All offspring have a 50 per cent chance of inheriting the dominant gene. Recessive traits are not normally passed to offspring when carried by only one parent. However, when both parents carry a gene for a recessive trait, such as red hair color, there is a 25 per cent chance that a son or daughter will have red hair.

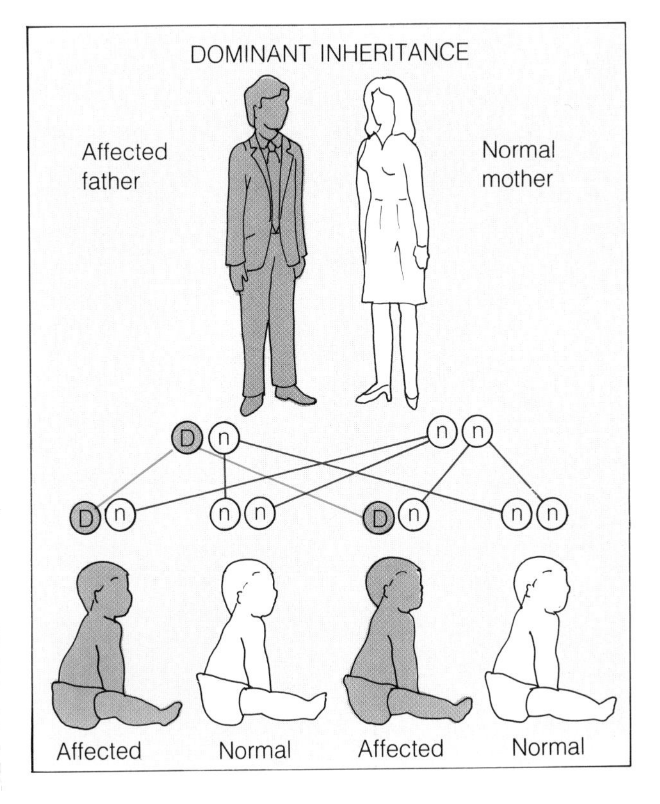

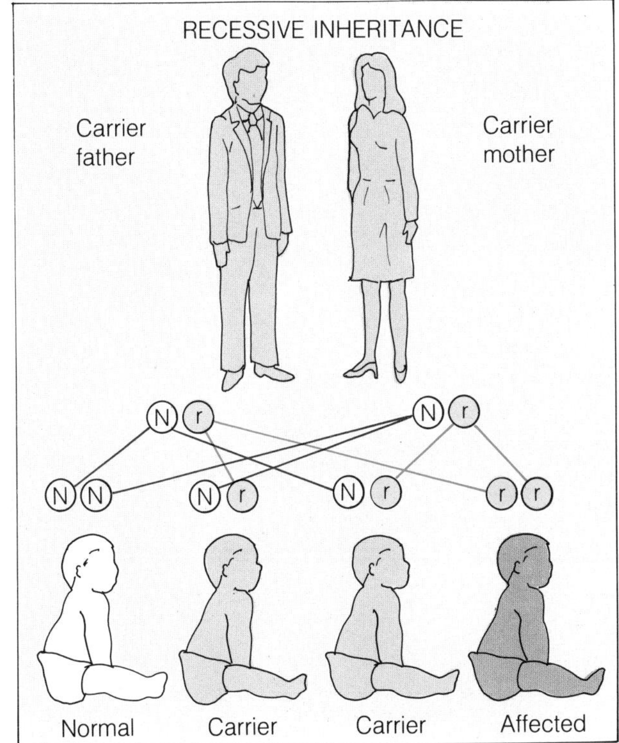

Examples of recessive inheritance of normal human features are uncommon. Red hair and very light blond hair both appear to be true recessive traits. It is often said that blue eyes are recessive to brown eyes, but this is not correct. It is an example of polygenic inheritance, which will be considered in more detail later in this chapter.

Another pattern of inheritance is known as codominance, of which blood grouping is a good example. The major blood group gene has three possible forms called A, B and O. Alternative forms of such a gene are called alleles. Even though a particular gene can have many alternative forms, any one person can have only two of them. With blood groups, the genes for group A and B are codominant — so that someone who inherits both an A gene and a B gene has group AB blood. The group O gene is recessive, and can produce group O blood only if a person has two O genes. This does not, however, prevent a parent with group A blood conceiving an offspring with group O blood, even if the partner has group B blood. Each parent may carry a "silent" group O gene which they pass on.

INHERITANCE OF HUMAN BLOOD GROUPS

Possible Gene Combinations	Results in	Blood Group
I^AI^A or I^AI^O	Antigen A only	A
I^BI^B or I^BI^O	Antigen B only	B
I^AI^B	Antigens A and B	AB
I^OI^O	No antigens	O

Human blood groups, on the ABO system, are defined in terms of antigens produced under genetic control and inherited in a predictable way. The gene concerned has a single chromosome location, termed the I locus. The relevant chromosome in an egg or sperm carries one of three genes: I^A, I^B or I^O. At fertilization, the genes pair in one of six possible ways, as in the above table, to produce the four common blood groups.

In the formation of sperm (left) *and egg cells* (right) *the usual 46 chromosomes per cell are reduced to 23. When the egg and sperm cells fuse at fertilization they form one cell that again contains 46 chromosomes.*

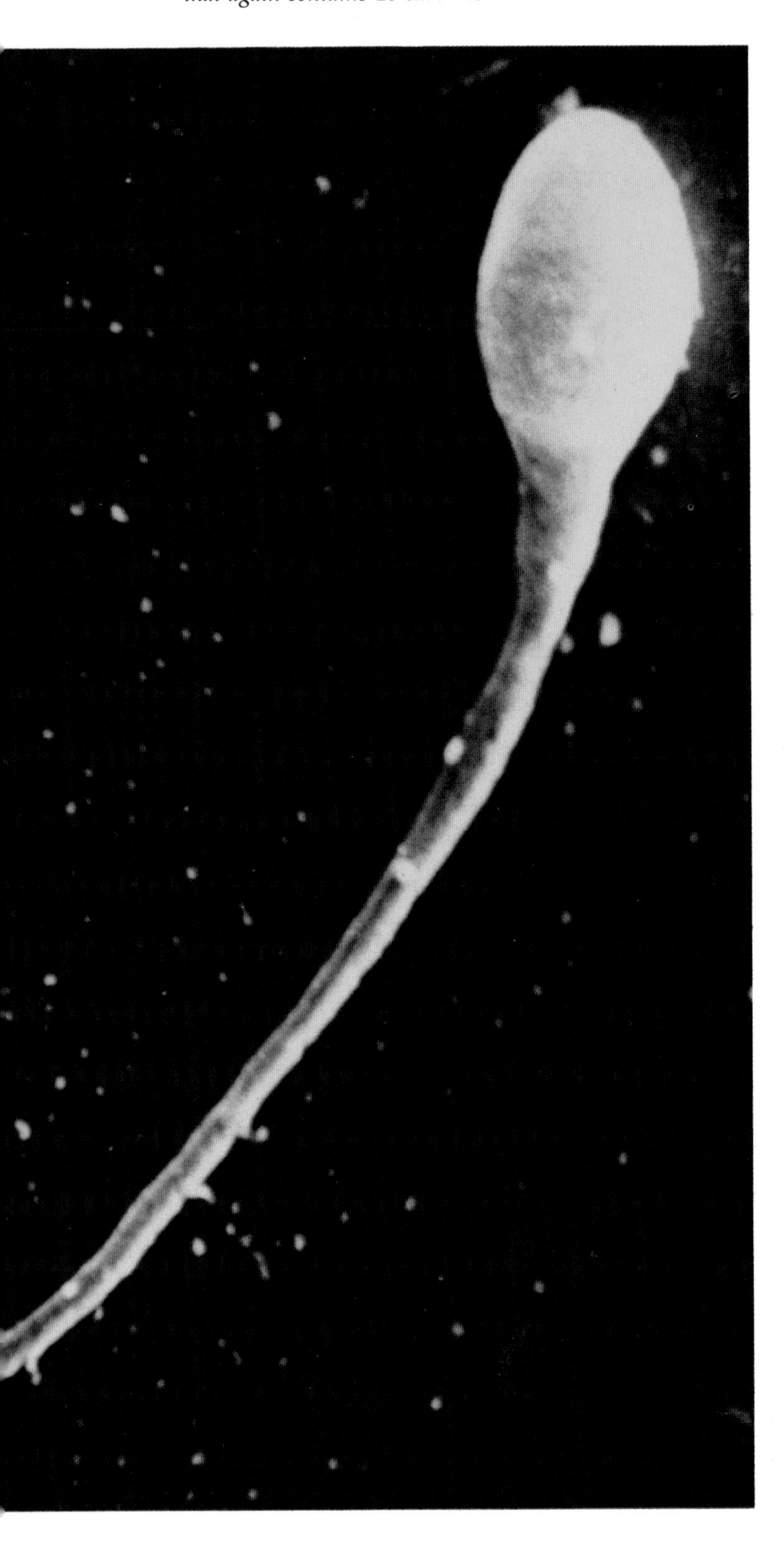

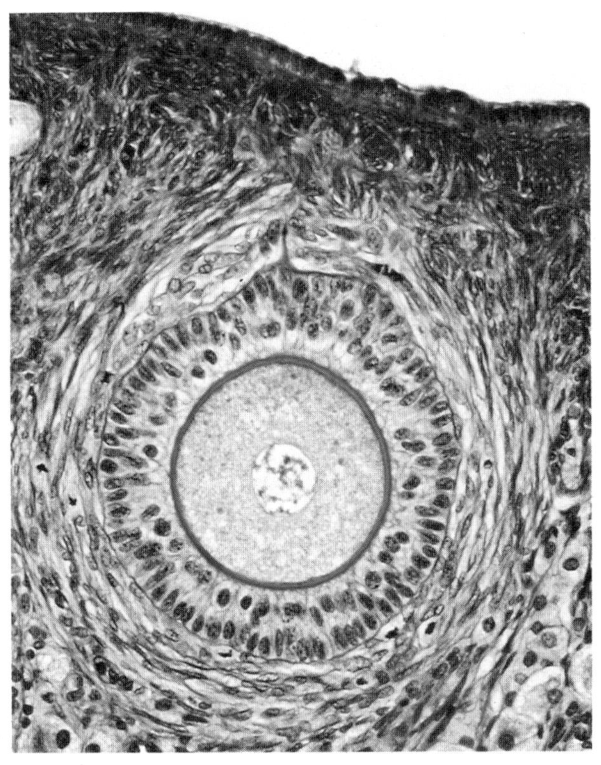

Blood group genetics may be used to determine the likely father in some cases of disputed paternity. A group B woman cannot have a group A child by a group O man, because he has only two recessive group O genes to pass to a child. Similarly a group AB man cannot have group O offspring, because he has only codominant A and B genes and one of these must be expressed in his children.

The Sex Chromosomes

An exception to the rule that there are "two genes in each pair, one from each parent" is the inheritance of genes on the sex chromosomes. Chromosomes are rodlike structures which can be seen in the nucleus of a dividing cell. Early in this century scientists realized that the way identical pairs of chromosomes separate in dividing cells is very similar to the way genes are shared out. Further weight was added to this observation when they studied germ cells in eggs and sperms, and noted that they have only half the number of chromosomes found in other cells. In fact only one of each pair of chromosomes was present. Fertiliza-

This illustration is an example of the tests used to detect color blindness. People with normal color vision see the number 15, whereas those with the sex-linked trait red-green color blindness see no number.

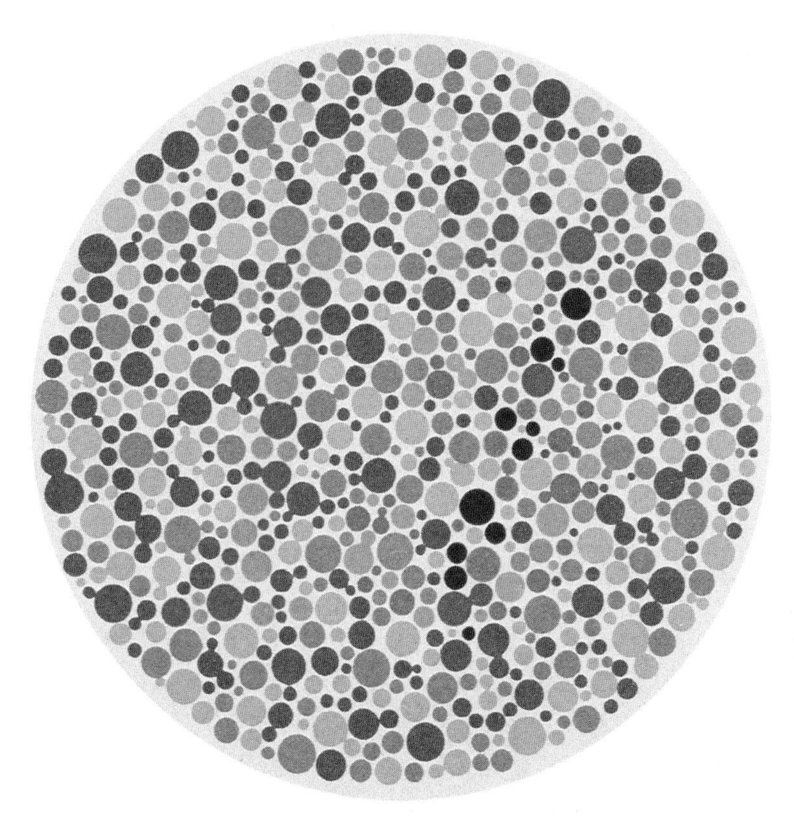

tion brings one half set of chromosomes from each parent together to form a complete set in the progeny. These observations constituted the chromosome theory of inheritance, which stated that genes must somehow be physically associated with chromosomes present in the nucleus. The behavior of the sex chromosomes and certain sex-linked traits provided support for this theory.

The American geneticist Thomas Hunt Morgan was working with fruit flies and noticed that a white-eyed fly had appeared among the normally red-eyed flies. Breeding this fly and others like it revealed a curious pattern of inheritance. The gene for white eyes seemed to be recessive in females but codominant in males. It was eventually proved that this resulted because one pair of choromosomes differed between males and females. We now know that females have two identical X chromosomes, whereas males have only one X chromosome; the male pair is completed by a different (Y) chromosome. A recessive gene on the X chromosome is blotted out by the presence of a dominant gene on the other X chromosome in females, but in males, with only one X chromosome, any recessive trait may become apparent through having no corresponding gene on the shorter Y chromosome.

There are two well-known sex-linked inherited abnormalities in humans — red-green color blindness and hemophilia.

Color blindness illustrates a puzzling property of sex-lined traits — their ability to "skip" a generation. Color blindness is not life-threatening, so color-blind men have a normal life span and produce normal numbers of children, passing their faulty gene on in the usual way. Because a color-blind father passes the X chromosome with the abnormal, recessive gene only to his daughters, all of his children have normal color vision because their mother's normal X chromosome masks its effect in the daughters, and the sons get their one X chromosome from the normal mother as well. But all the daughters are carriers, and their sons may inherit the X chromosome with the color-blind gene and be affected. Grandfathers and grandsons are color blind, but the trait skips the intervening generation. (It is possible for a woman to be born color-blind too if she is unlucky enough to inherit recessive genes on the X chromosome from both of her parents.)

Hemophilia-A is a bleeding disorder caused by the absence of a protein required for blood clotting. The gene that governs the production of this protein, factor VIII, is on the X chromosome. A woman has two factor VIII genes and, therefore, if one gene is abnormal the other produces enough clotting factor to prevent excessive bleeding. A man, with only the one X chromosome, has little or no clotting factor if this gene is faulty, because there is no second gene to take its place. So a woman carrying the abnormal gene is unaffected — but passes the trait to her male offspring.

The Royal Disease

The English Queen Victoria was a carrier of hemophilia, and at the turn of this century the disease was regarded as just another illness sent to plague the royal parents of Europe. Queen Victoria's son Leopold had hemophilia and died when he was thirty-two years old after a minor blow to the head. Two of her daughters were

The blood clotting disorder hemophilia displays simple sex-linked inheritance which can be illustrated by tracing the line of carriers and sufferers in descendants of the English Queen Victoria. The hemophilia trait can be transmitted in several different ways; the main ones are shown in the yellow circles. In (A), the defective gene is carried on one of the mother's X chromosomes. Disease follows when the X chromosome containing the defective gene is transmitted to a son. There is a 50 per cent chance that sons will be affected, and a 50 per cent chance that daughters will be carriers. Inset (B) shows a

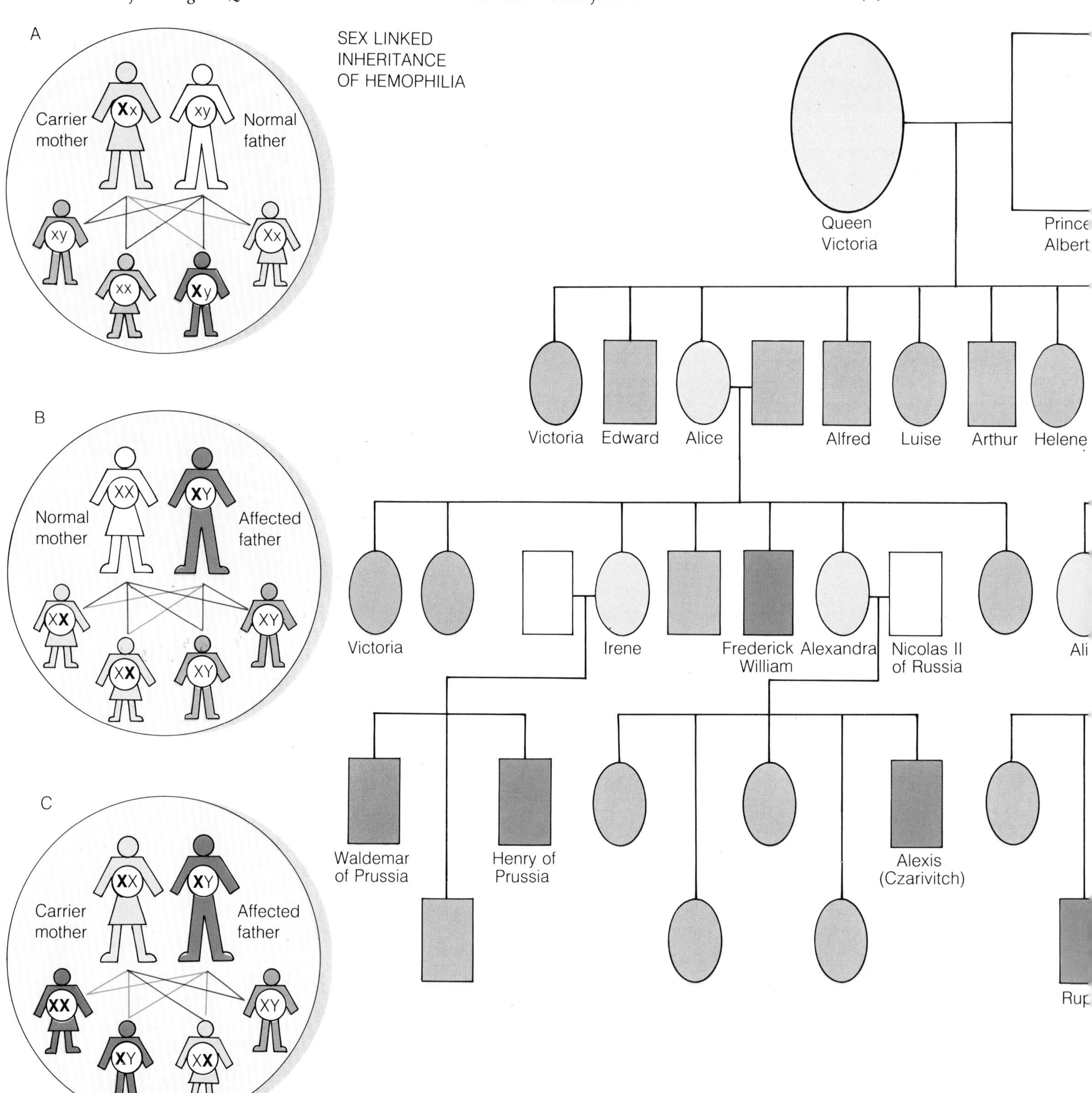

hemophilic father and a normal mother. The father transmits his X chromosome, carrying the hemophilia gene, to his daughters who almost certainly become carriers, while sons remain unaffected. In (C), both parents possess a defective gene on an X chromosome. Daughters will either be affected by, or carriers of, the disease, while there is a 50 per cent chance that sons will not inherit it.

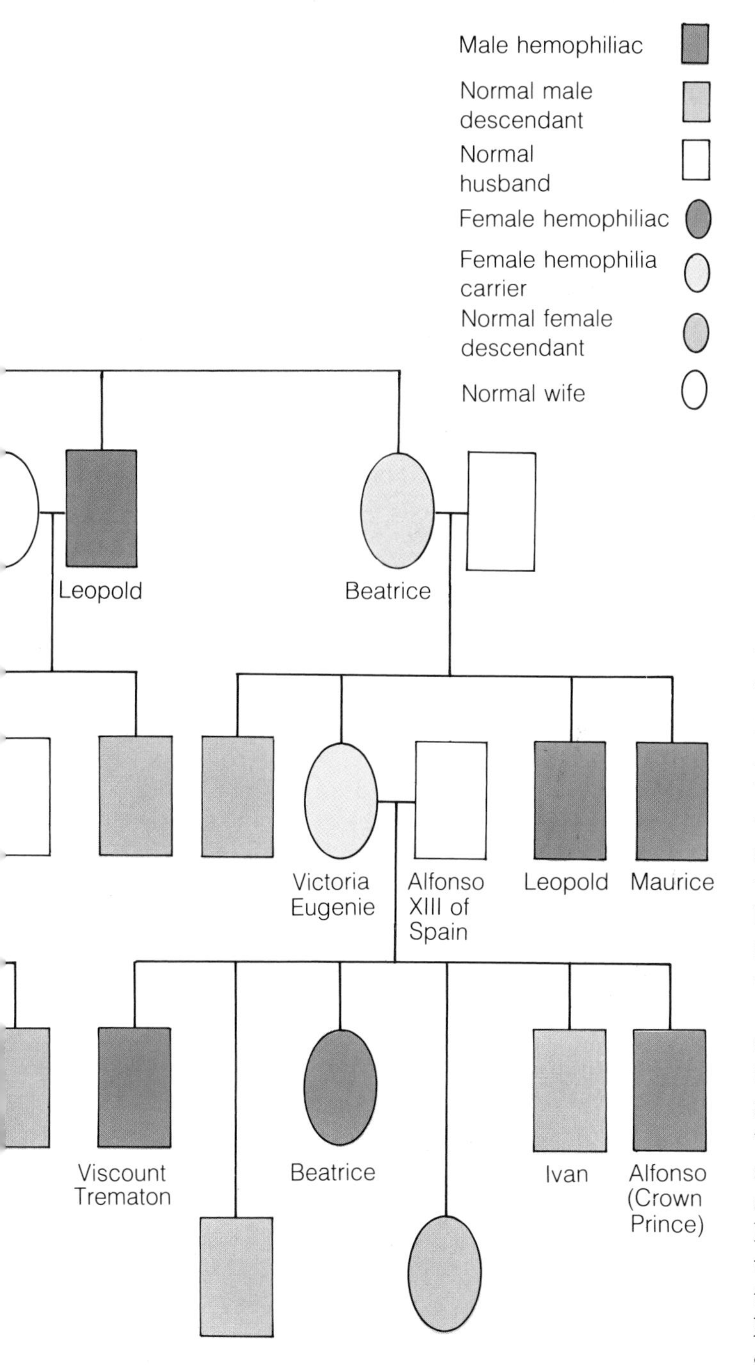

carriers and, through their descendants, took the disease to the royal houses of Spain, Germany and Russia. In Spain the young princes lived in padded apartments, and the trees in the park where they played were also padded to reduce the risk of injury to the children.

But by far the most devastating result of this royal disease was the fall of Russia. It may seem a sweeping statement to say that the affliction of one small prince resulted in the downfall of this powerful Empire, but even at the time key figures in the drama recognized this essential factor. Alexander Kerensky, who became Premier of Russia, said "If there had been no Rasputin, there would have been no Lenin," and the geneticist John Haldane said "Rasputin took the empire by stopping the bleeding of the Czarevitch."

It is interesting to contemplate how the evil Rasputin gained so much power in the Russian court. Today hemophiliac children can be helped by transfusions containing factor VIII, and pain can be relieved with drugs. But the young Czarevitch, Prince Alexis, had no such help. Imagine the anguish that his mother Alexandra had to bear watching her precious son, the first male heir born to a reigning Czar since the seventeenth century, suffer the agonies of hemorrhage with no relief for the pain except fainting. During bouts of internal bleeding Alexandra would sit with him day and night; doctors were powerless to alleviate his terrible suffering.

In 1912, when the family was in Poland, Alexis developed a massive hemorrhage in his hip and thigh. For days the blood flowed internally with nowhere to escape, the seven-year-old boy was wracked with the most terrible pain, and the servants in the house stopped their ears against the screams. Eleven days passed, hope failed and all Russia prepared for the news that the Czarevitch was dead. In desperation Alexandra sent a telegram to Rasputin, a wandering priest whose healing powers were famous. Rasputin telegraphed back "God has heard your prayers. Do not grieve. The little one will not die. Do not allow the doctors to bother him too much." A day later the hemorrhage stopped and Alexandra's faith in Rasputin was complete and unshakable.

In the years that followed, leading up to the

The English Queen Victoria was a carrier of the hemophilia trait. She bore a son who suffered from the disease, and the disorder appeared among the male descendants of two of her daughters.

Russian Revolution, it was Rasputin who — through Alexandra — stopped any attempts by Czar Nicholas to share out Imperial power. Without Rasputin the Czar would have undoubtedly lost absolute power, but that might have been replaced by a constitutional monarchy similar to the one in England under the Czar's cousin, King George V. Instead it resulted in the massacre of the Czar, his wife, their four daughters, Alexis and four faithful family servants in a basement of Ekaterinburg in the Urals district of Siberia.

The Laws of Chance

There is no preordained way to determine heredity; it is governed by the laws of chance and probability. An egg may be fertilized by a sperm containing an X chromosome and become a female embryo, or a sperm containing a Y chromosome and become a male embryo; either could happen.

The mathematics governing the inheritance of sex is essentially the same as the mathematics of tossing a coin. Half the time the coin lands heads and half the time tails. Try tossing a coin and keep a note of the results. You will probably find that you toss heads or tails several times in a row at some time during the game. In general, you will toss heads twice in a row once in four times, and three times once in eight. This is "predicted" by the laws of chance, which say that if something occurs by chance half of the time then it occurs twice a quarter of the time ($\frac{1}{2} \times \frac{1}{2}$), or an eighth of a time ($\frac{1}{2} \times \frac{1}{2} \times \frac{1}{2}$) three times in a row.

The same thing happens with children. On average one family in eight with three children has three boys or three girls, and one family in thirty-two with five children has five boys or five girls. Notice, however, that five boys or five girls both occur with the same probability, so that one family in sixteen with five children has children of all the same sex. And since such families balance out the numbers of boys and girls, the overall result is an equal number of each sex.

Although the theoretical inheritance of X or Y chromosomes and hence the sex of children is random, reality is actually a little different. About 106 male children are born for every 100 females. No one knows why this is so, although one ingenious suggestion is that sperm carrying the smaller Y chromosome are faster than sperm carrying an X chromosome and therefore reach and fertilize the egg slightly more often. Whatever the reason, the numbers of both sexes are about equal within a few years of birth because more male than female children die within the first few years of life.

The laws of chance also show that certain rare events such as having ten daughters will definitely occur if the population is large enough. Indeed, one family in a thousand with ten children has all girls. Any particular combination of sexes can occur, and the laws of chance give the odds or probability of a particular result for any particular couple.

Polygenic Inheritance

The laws of chance govern the inheritance of all genes because each pair of chromosomes is sorted at random when packaged into eggs and sperm — a principle called "independent assortment." For simplicity it can be assumed that independent

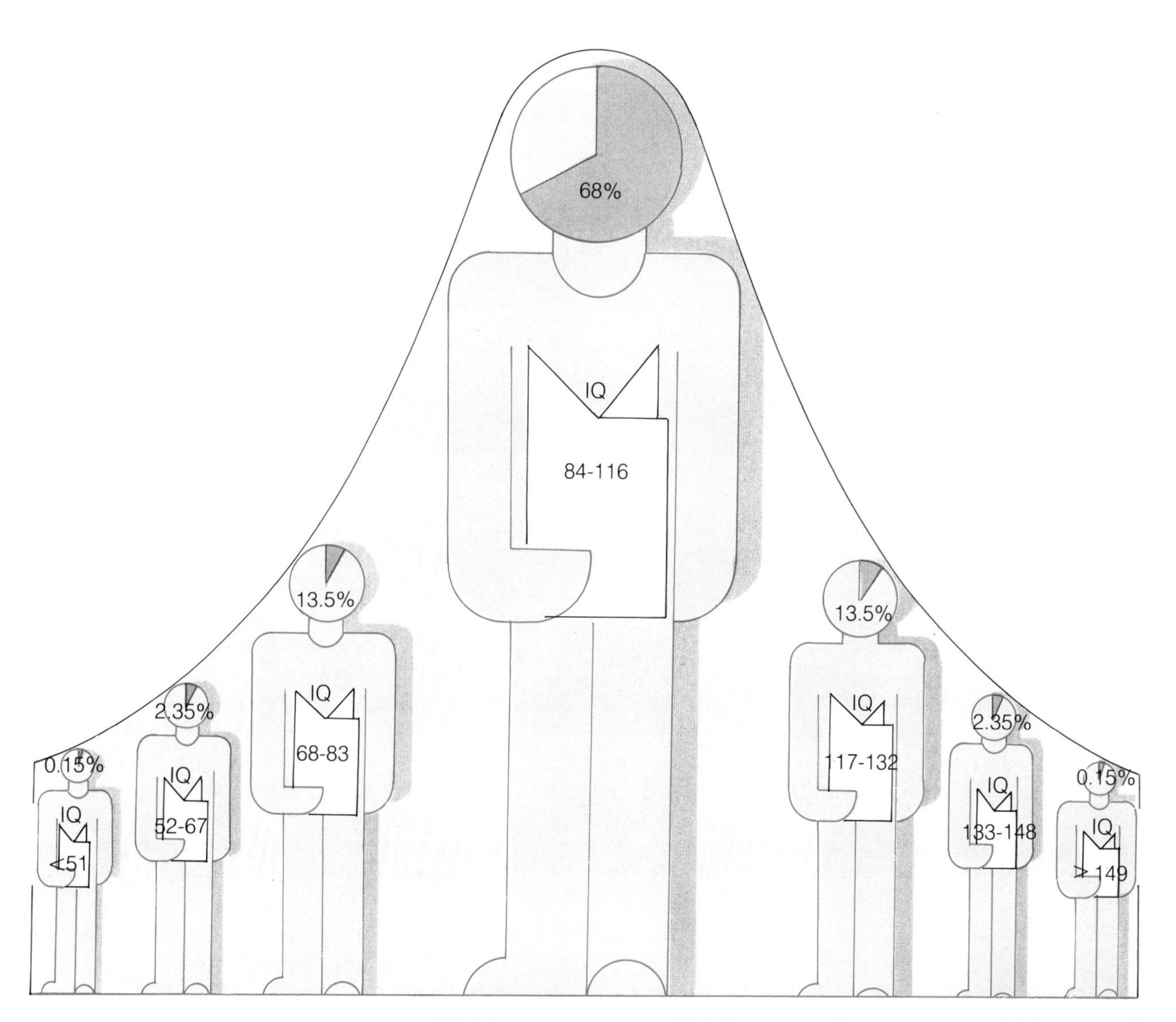

When large amounts of data such as intelligence test scores (above) *are plotted on a graph, they often fall into a symmetrical bell shape known as the normal distribution. Most scores fall near the mean (100), thus giving the highest point of the bell. Figures taper off sharply at very high and very low scores. This curve also arises when the outcome of a process is based on a large number of chance events that all occur independently. The laws of chance govern the inheritance of sex. The chance of having a large number of children with equal numbers of girls and boys* (right) *is thus extremely low.*

J. B. S. Haldane

Mathematician Among Geneticists

Much of John Burdon Sanderson Haldane's early research was on a subject we participate in all the time without thinking about it—breathing. Nevertheless, Haldane is now famous not only for pioneering work in physiology but also for his amazing insights into genetics and—crucially—the interdisciplinary knowledge to combine with it to be able to derive important mathematical conclusions.

Haldane was born in Oxford, England, in November 1892. From school at Eton College he went to New College, Oxford University, where he graduated in mathematics but also attained excellent results in philosophy and the Classics. World War I then broke out and he served on the Western Front in the Black Watch Regiment until the war ended. After a few years more in Oxford, he moved to take up the position of Reader in Biochemistry at Cambridge University. Over the next decade he additionally became Supervisor of Genetical Research at the John Innes Horticultural Research Station at Merton (1927), and Fullerian Professor of Physiology at the Royal Institution (1930). This triple situation changed when in 1933 he took up the post of Professor of Genetics at

University College, London, four years later transferring at the same establishment to the chair of Biometry.

Haldane had been interested in genetics from about the age of nine. (Later he noted hereditary effects among his sister's huge menagerie of guinea pigs.) Mendel's work had just been rediscovered, and he set himself the task of establishing quantitative laws governing the mathematical elements in the theory of natural selection and the effects of heredity. What became known as Haldane's Law followed as a result in 1922. Although he went on to investigate the structure of genes, particularly the composition of the X chromosome, his next most significant work was the authoritative publication *Enzymes* (1930).

At University College he worked to link mathematics to the rates of genetic mutation, and in 1936 discovered the hereditary resemblance between hemophilia and color blindness.

But by this time Haldane had become a convinced and outspoken Marxist; he joined the Communist Party at the outbreak of World War II. His war effort was to research once more into respiration. Even while at school he had helped his famous physiologist father, John Scott Haldane, in his investigation into respiration with special reference to safety in mines and for deep-sea divers. Wounded twice as a young soldier during World War I, he returned to his native England not to convalesce but to rejoin his father in research.

After the war, in the 1950s, however, he became disillusioned both with the Communist Party (the Soviet Union afforded what he considered excessive credibility to the biologist Trofim Lysenko) and with the British Government (who controversially invaded Suez at that time). After emigrating to India, he became a naturalized citizen of that country in 1961. He died there in 1964.

Weird but beautiful, this is the lionhead veiltail, one of more than a hundred species of mutated goldfish. This fish's "abnormalities" consist of a three-lobed flowing tail and a swollen "hood" on its head.

assortment applies to all genes (although in reality there is a common exception called linkage, which occurs when two genes are close together on a chromosome and are carried around together).

Most human characteristics, such as height, are determined by several genes acting together. It is one example of polygenic inheritance. When a number of genes influence a trait, its inheritance becomes complicated and almost totally unpredictable. For example, the famous basketball player Wilt Chamberlain is about seven feet three inches tall but his parents were less than five foot nine.

The laws of chance come into play here. Assume that there are only two genes — say AA and BB — that determine how tall a person is. Assume also that each gene has two forms (alleles) A and a or B and b. A and B genes make a person taller and a and b genes make a person shorter. An average-sized person has one of each, with the genetic makeup AAaa BBbb. When two such average people have children, the genes are sorted at random. Each parent has sex cells (called gametes) which contain AABB, AAbb, aaBB or aabb genes. They are equally likely, and when they fertilize each other the chances of any one egg meeting any one sperm are equal. So a AABB egg is fertilized by a AABB sperm once in sixteen times ($\frac{1}{4} \times \frac{1}{4}$, the chance of AABB occurring in either gamete). And the same is true for the genes for being shorter than average; one child in 16 inherits all four shortness alleles.

Therefore "average" people can have far from average offspring, and they can vary in either direction. Overall, however, "average" parents have average-sized children just under half the time (six out of sixteen).

When many genes each add or subtract a small amount to the total, the result is a continuous range of variation. Tall parents usually have children who are taller than average, but not as tall as their parents. Similarly for short parents: some of their children are nearer average height than they are. This phenomenon is called "regression to the mean." Children tend to be more "average" than their parents. Regression to the mean appears to operate in connection with many other human characteristics, including intelligence.

The laws of chance and their relative statistical probability are used by geneticists to investigate how a particular trait is inherited. By studying the outcome of many different matings it is possible to decide if a particular feature results from dominant, recessive, sex-linked or polygenic inheritance. Similar analysis is used when trying to determine how a particular human disorder is inherited. Genetic counseling largely depends on an understanding of these principles.

Mutations

Genes are not unchangeable units of inheritance. They can change, or be changed by environmental influences. A change in a gene is a mutation.

Mutations were originally noticed in domestic animals such as cats and dogs, in which occasional unusual offspring called "sports" appeared. Some of the sports were kept and used for breeding, leading to the large numbers of different breeds of dogs and cats which now exist. Similar events led to the development of the various breeds of most farm animals. An animal with a new desirable

The tumors on this fruit fly's head are an example of changes that may occur as mutations. The imperfect self-reproduction of chromosomal DNA may produce the potential for new traits.

This train crash was set up to assess the suitability of containers proposed for use in the transportation of radioactive materials. Radiation is a powerful mutagen and exposure to it is likely to cause genetic changes.

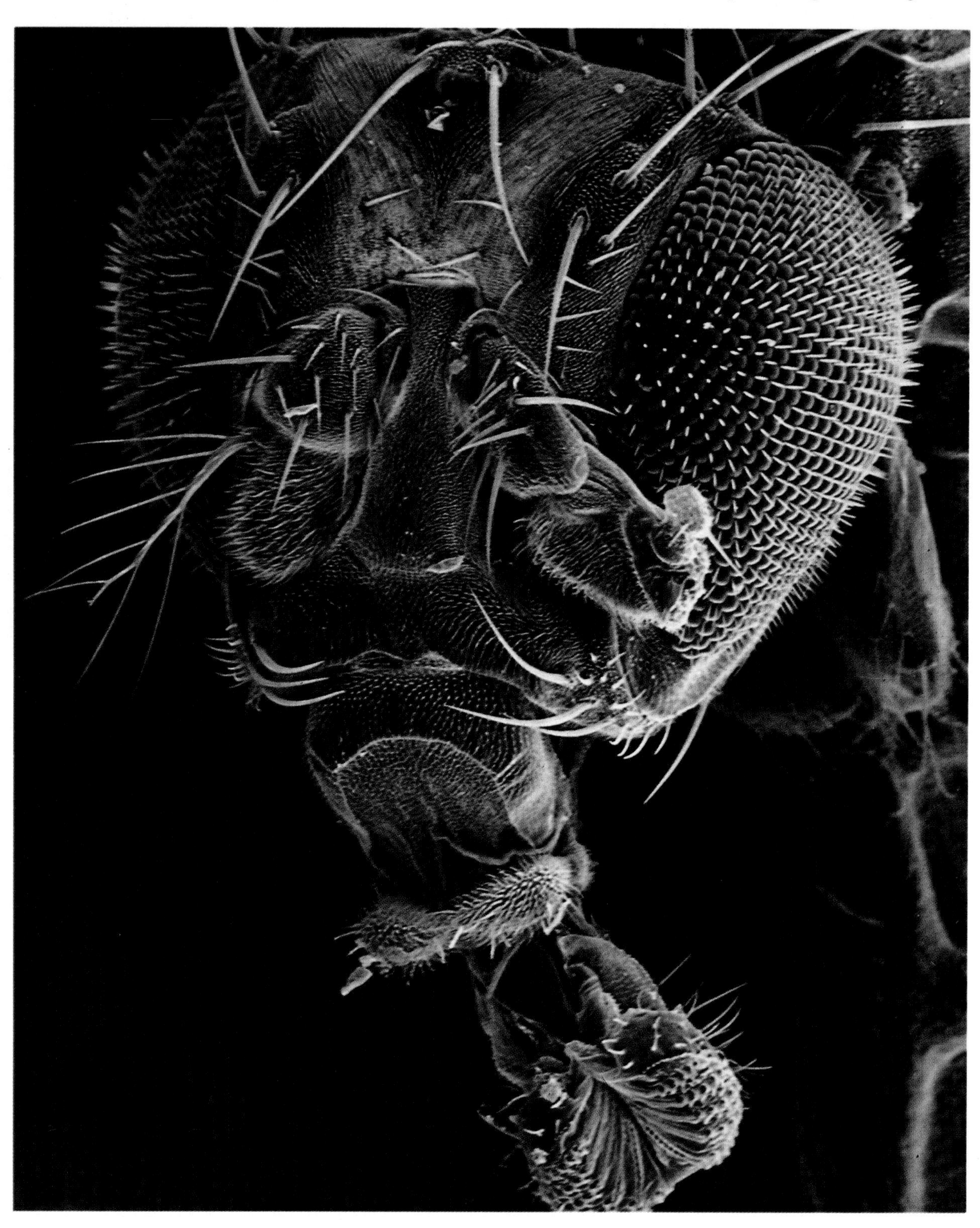

feature that appears by chance — for example long straight wool — is used to produce a new strain of animal which retains the desirable characteristic. Arguably all agriculture is based on mutation and selection of plants and animals.

Although mutations have been occurring for millions of years, Hermann Muller, the American geneticist, was the first to develop experiments to study how and why they occurred. The results have far-reaching implications for humanity, because his experiments showed that mutations could be induced by radiation. At the time, in the 1920s, radiation from X rays or radioactive elements was an insignificant hazard for most people, and was regarded merely as a curiosity. Today radioactive pollution and atomic weapons make radiation damage a great concern to everyone.

Muller's experiments also established another worrying aspect of mutations — most are harmful. Muller set about trying to find out how genetic damage could be avoided and what agents caused the damage.

Mutation and mutagens (substances that cause mutation) are significant in two distinct but related ways. One is the effect on offspring caused by damage to the egg- and sperm-forming cells in the parents. For example, it has been suggested that one of the chemicals used to kill vegetation in Vietnam, Agent Orange, may have damaged the genetic "apparatus" of individuals exposed to it. Claims that the victims have produced deformed children years after exposure to the chemical are still being tested in the courts.

Radiation may produce inherited mutations, but this has been difficult to prove in human beings. So far no increase in mutations has been found among the descendants of people who were present at the atom bomb explosions at Hiroshima and Nagasaki. It may be that the number of children at risk is too small for the effects of relatively rare events to be detected. Mutations are subject to the laws of chance, and even if the number of mutations were increased from one in a million to one in 10,000 (a 100-fold increase), no increase would be found if only 1,000 children were investigated. Large numbers of people are needed to study large changes in the frequency of rare events. For this reason, new and potentially mutagenic drugs have

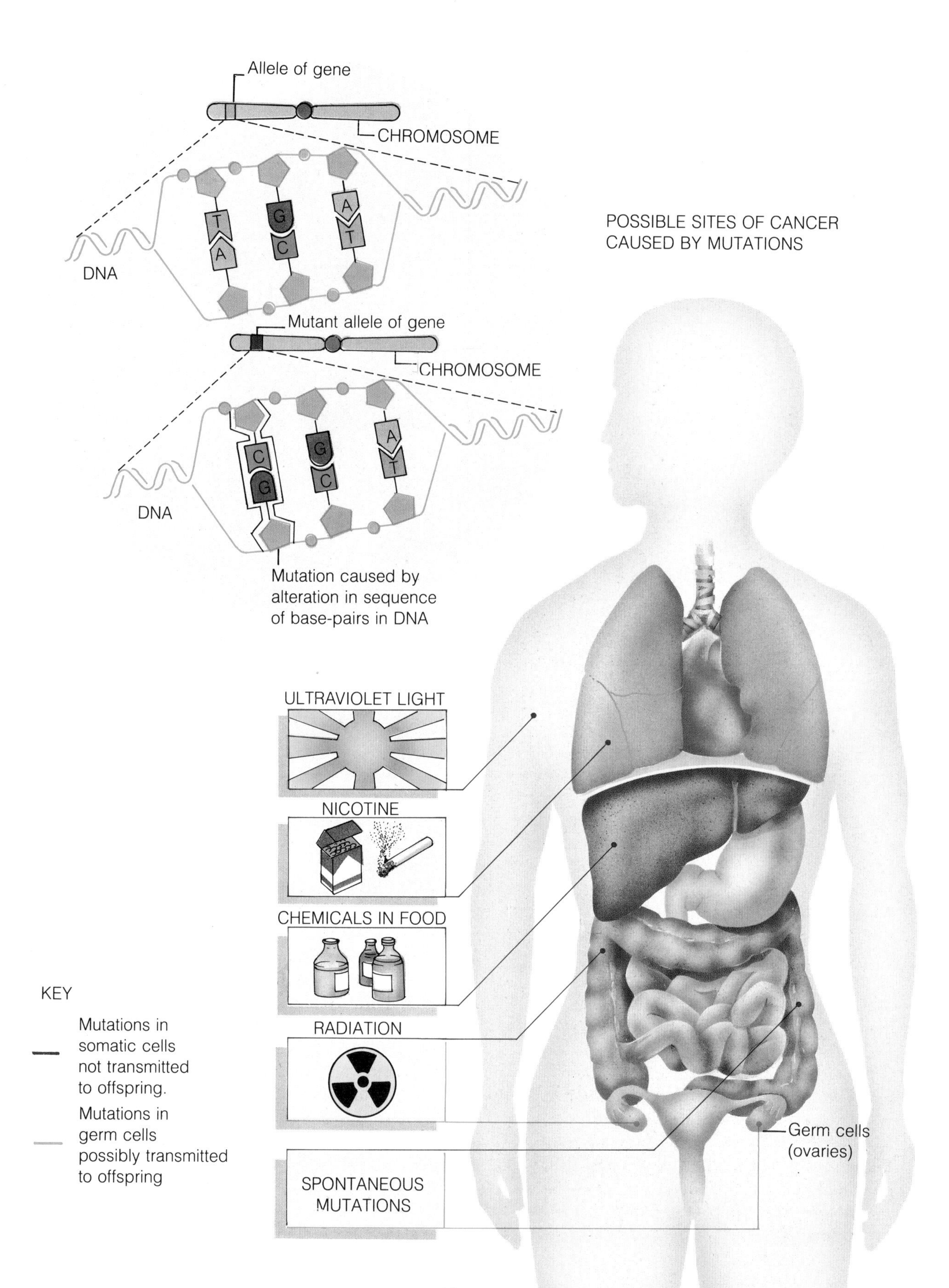
Allele of gene
CHROMOSOME
T
A
G
C
A
T
DNA
Mutant allele of gene
CHROMOSOME
C
G
G
C
A
T
DNA
Mutation caused by alteration in sequence of base-pairs in DNA
POSSIBLE SITES OF CANCER CAUSED BY MUTATIONS
ULTRAVIOLET LIGHT
NICOTINE
CHEMICALS IN FOOD
RADIATION
SPONTANEOUS MUTATIONS
Germ cells (ovaries)
KEY
Mutations in somatic cells not transmitted to offspring.
Mutations in germ cells possibly transmitted to offspring

Mutations occur mainly as a result of changes in the single base-pairs that comprise the DNA. Only mutations occurring in the germ cells (egg and sperm) are likely to be passed to offspring.

to be tested on large numbers of animals to make sure that there is no possibility of a genetic disaster.

The second area in which mutation and mutagens affect humans directly occurs when body cells other than sex cells are damaged. So far heredity has been considered only as it affects inheritance from one generation to the next. It must not be forgotten that our bodies are replenishing themselves by forming new cells to replace damaged or worn out cells all the time. New cells are usually exact copies of the ones they replace, because they have the same genes and the same environment. But a mutation can lead to dramatic and unexpected changes.

Yet spontaneous mutations among humans occur all the time. Whether they result from "accidents" — such as minor slips in reproducing the genetic code — or from inescapable environmental effects — such as natural radioactivity in rocks and cosmic rays — is often debated. Most mutations occur in the somatic (non-sex) cells, and therefore do not affect heredity. But a mutation in an egg- or sperm-forming cell can be passed on to future generations. The person carrying the mutation shows no sign of it because it is in the sex cell — only their descendants may suffer.

As with other genes, new mutations may be either dominant or recessive. Dominant changes probably affect any progeny receiving that gene. Most mutations are recessive and very few are wholly good. But despite their almost uniform deleterious effects, mutations are the essential elements for any change in the genetic constitution of a population—they make possible the mechanism of evolution. A recessive mutation does not show up until, by chance, a recipient has children by someone with a similar recessive gene. And even then there is only a one in four chance that they will have an affected child.

For most recessive genes the odds of mating with someone carrying the same abnormal gene are many thousands to one, assuming of course that the partner is chosen at random from a sufficiently large population. The odds change dramatically for the worse if the parents are blood relatives, especially first cousins. First cousins have grandparents in common so that if one partner carries the gene, then there is a one in eight chance that the other does as well. The chance falls if the two partners are more distant cousins, but is still substantially higher than for two totally unrelated people having the same abnormal gene.

A killer T cell seeks out and attacks an invading tumor cell. Hereditary disorders of the immune system sometimes result in a breakdown of this process, leaving the individual more susceptible to disease.

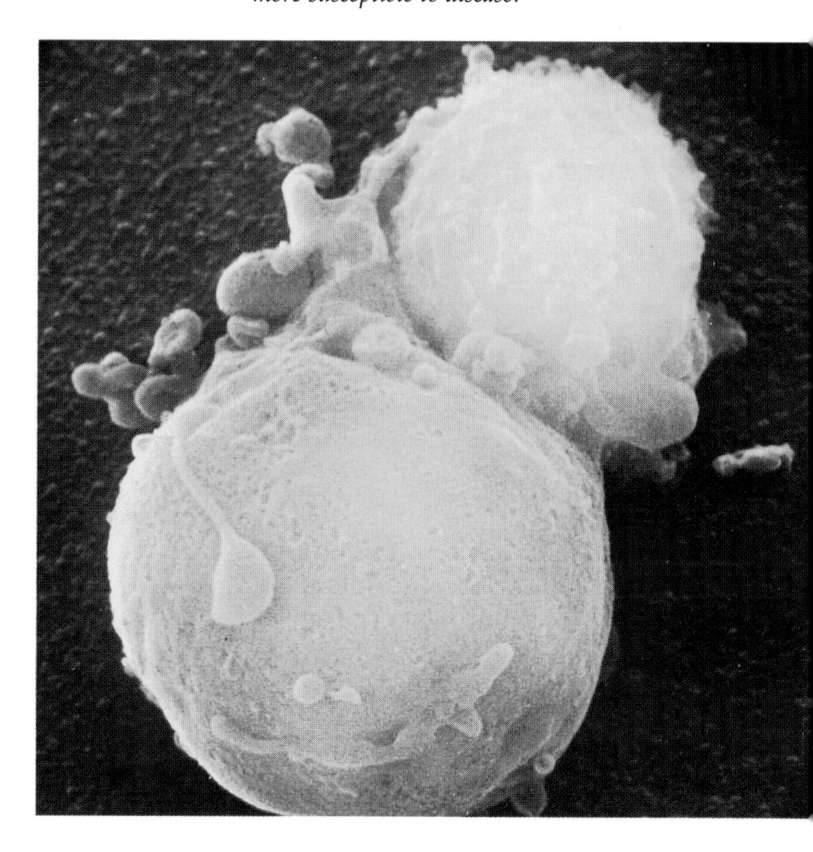

Cancer is thought to be caused by genetic damage to cells. Instead of inheriting exactly the right program for their place in the body, cancer cells have acquired — through mutation — the capacity for uncontrolled growth and sometimes the ability to spread through the body.

Substances that can cause such cancer-producing mutations are called carcinogens. They include chemicals and radioactive compounds, as well as naturally occurring environmental factors such as tobacco tars, asbestos and sunlight. The usual way to find out whether or not a given substance is safe is by testing its effect on some well-defined experimental model. This involves using large amounts of the chemical and testing it on large numbers of animals before interpreting the results. A quicker and cheaper test, called the Ames test (after its developer at the University of California),

A perusal through a family album often reveals a number of individuals with some kind of genetic resemblance. Tests can now be performed to assess the likelihood of passing potentially damaging genes to offspring.

is based on the knowledge that most carcinogens cause mutations. By determining the effect of the substance on the number of mutations in bacteria, the cancer-causing potential can be estimated.

Muller carried out experiments that called attention to the ability of radiation to produce genetic damage. We now know that one effect of such damage is an increase in the risk of certain kinds of cancer, especially cancer of the blood-forming cells (leukemias). Tragically this effect is produced not only by large doses of irradiation (for example after a nuclear explosion) but also by small doses. It has been shown, for example, that X-ray examination during pregnancy may increase the chance of leukemia in the exposed fetus. There is much debate at present as to whether there is any "safe" low dose of irradiation which does not cause damage to genetic material.

Many rare human recessive disorders are largely confined to populations in which interbreeding of close relatives is common. Isolated rural or island communities, such as those found in parts of the Appalachian mountains, certain religious sects such as the Amish, or self-selected subcultures like the Brandywine people of Maryland, are particularly likely to be affected by interbreeding. These communities are often descended from a small number of founders and tend to marry among themselves, so that everyone in the group is a blood relation to some extent. Studies on modern human populations, such as the people in and around Hiroshima and Nagasaki, have shown that marriages between first cousins are perhaps as much as twice as likely to produce fatally damaged children as unrelated marriage. This still holds true even in a population not exposed to atomic radiation.

Analyzing a Family Pedigree

How is human heredity studied? Human generations are long and genetic disorders rare, so what

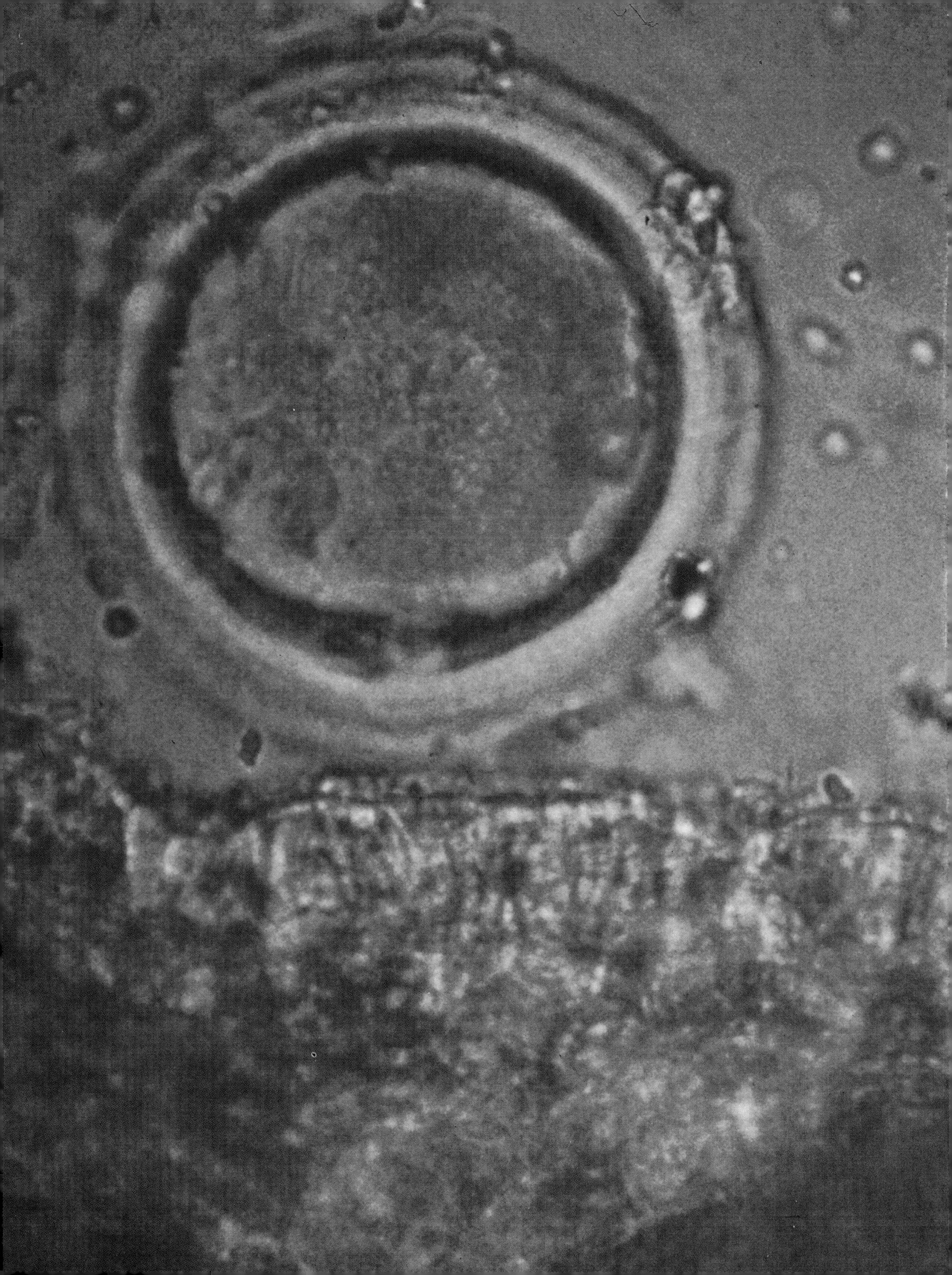

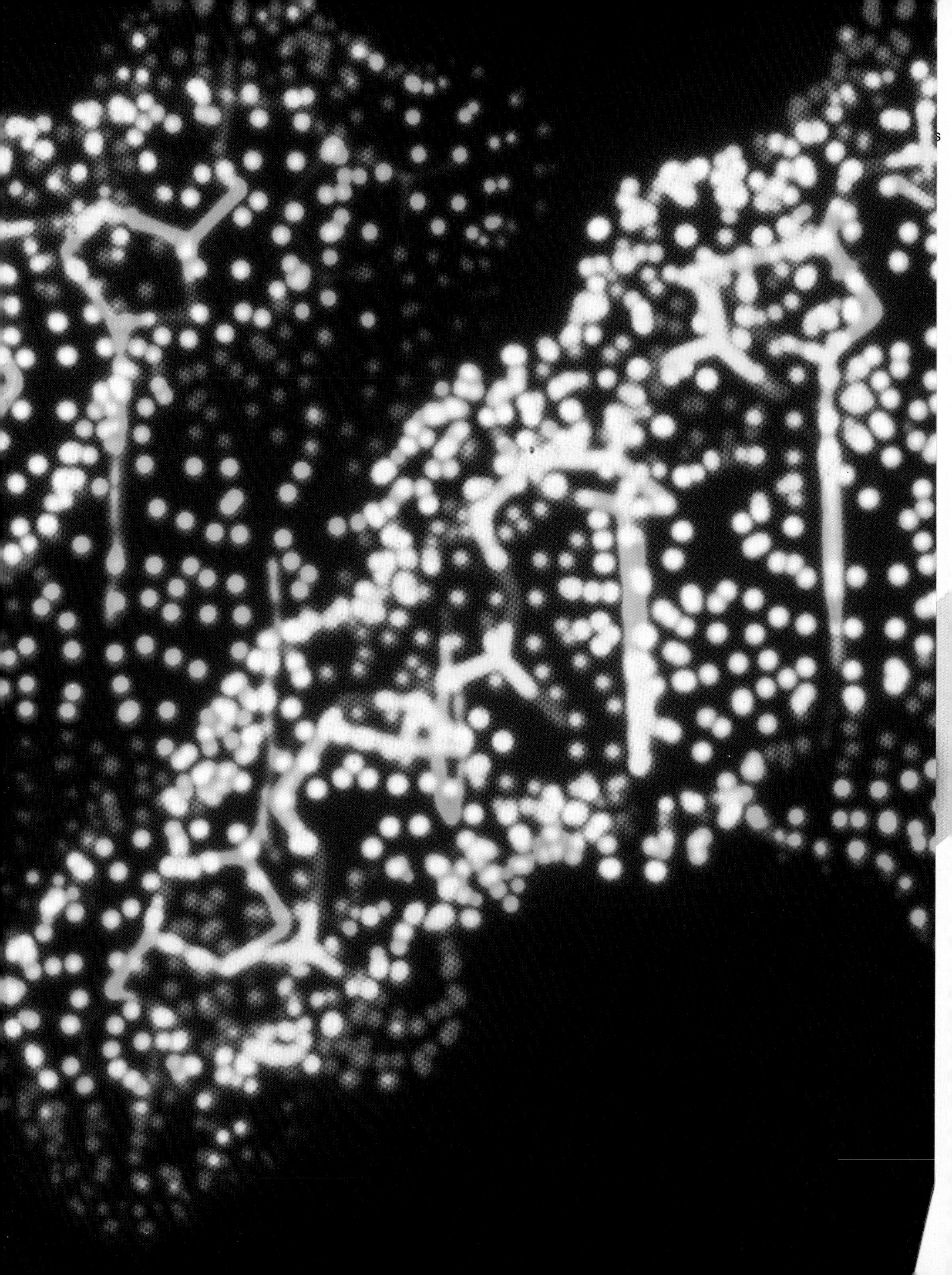

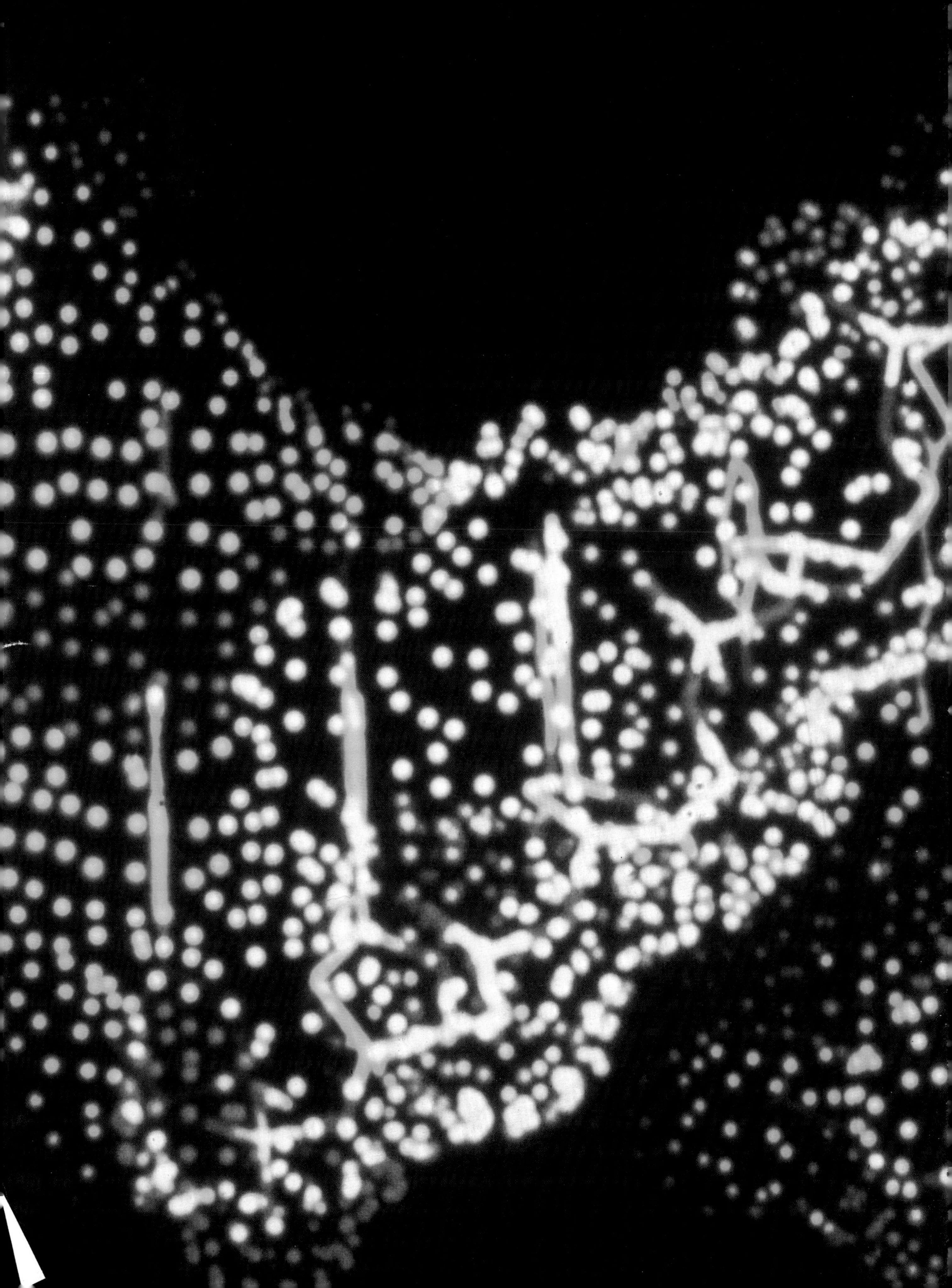

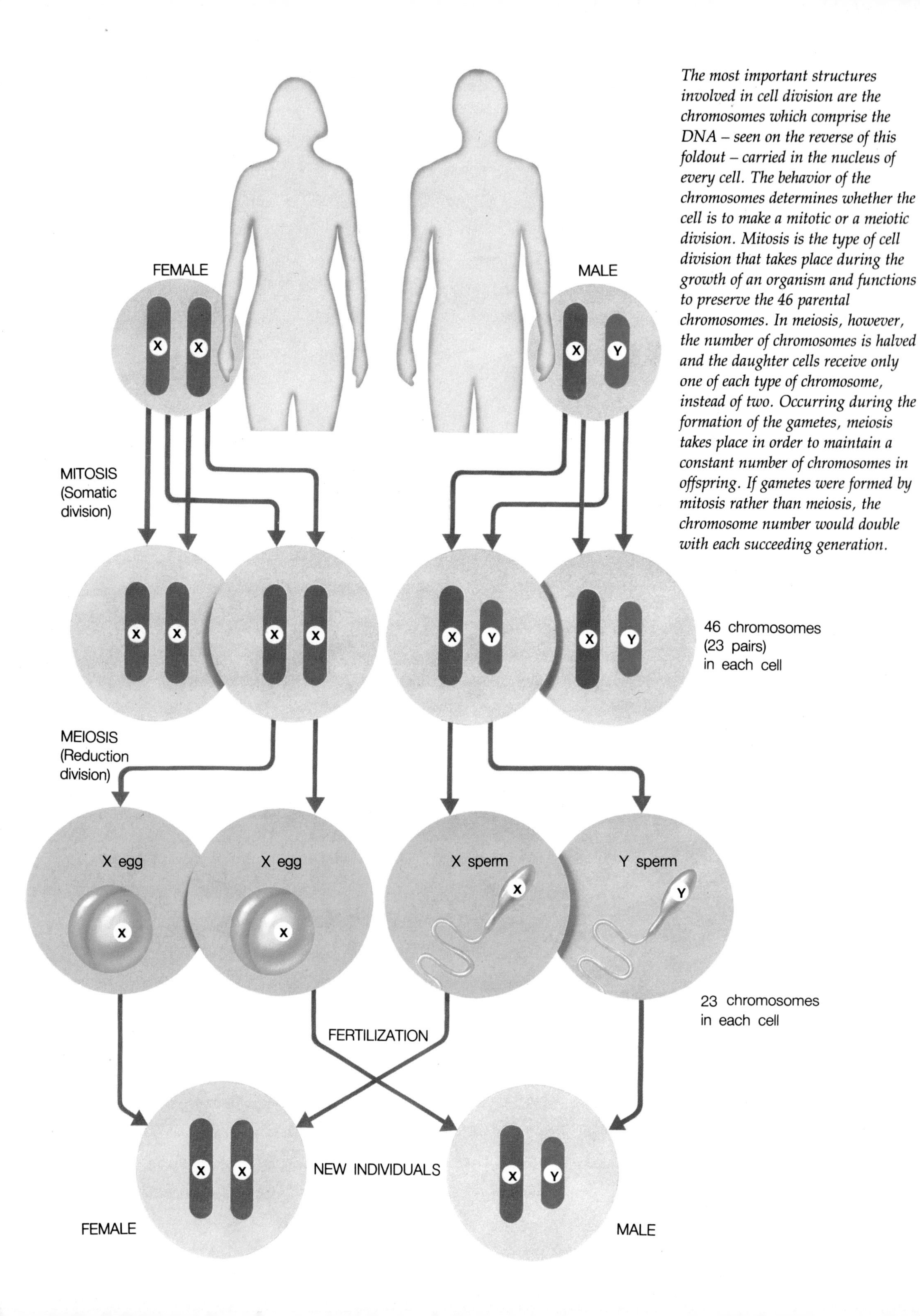

The most important structures involved in cell division are the chromosomes which comprise the DNA – seen on the reverse of this foldout – carried in the nucleus of every cell. The behavior of the chromosomes determines whether the cell is to make a mitotic or a meiotic division. Mitosis is the type of cell division that takes place during the growth of an organism and functions to preserve the 46 parental chromosomes. In meiosis, however, the number of chromosomes is halved and the daughter cells receive only one of each type of chromosome, instead of two. Occurring during the formation of the gametes, meiosis takes place in order to maintain a constant number of chromosomes in offspring. If gametes were formed by mitosis rather than meiosis, the chromosome number would double with each succeeding generation.

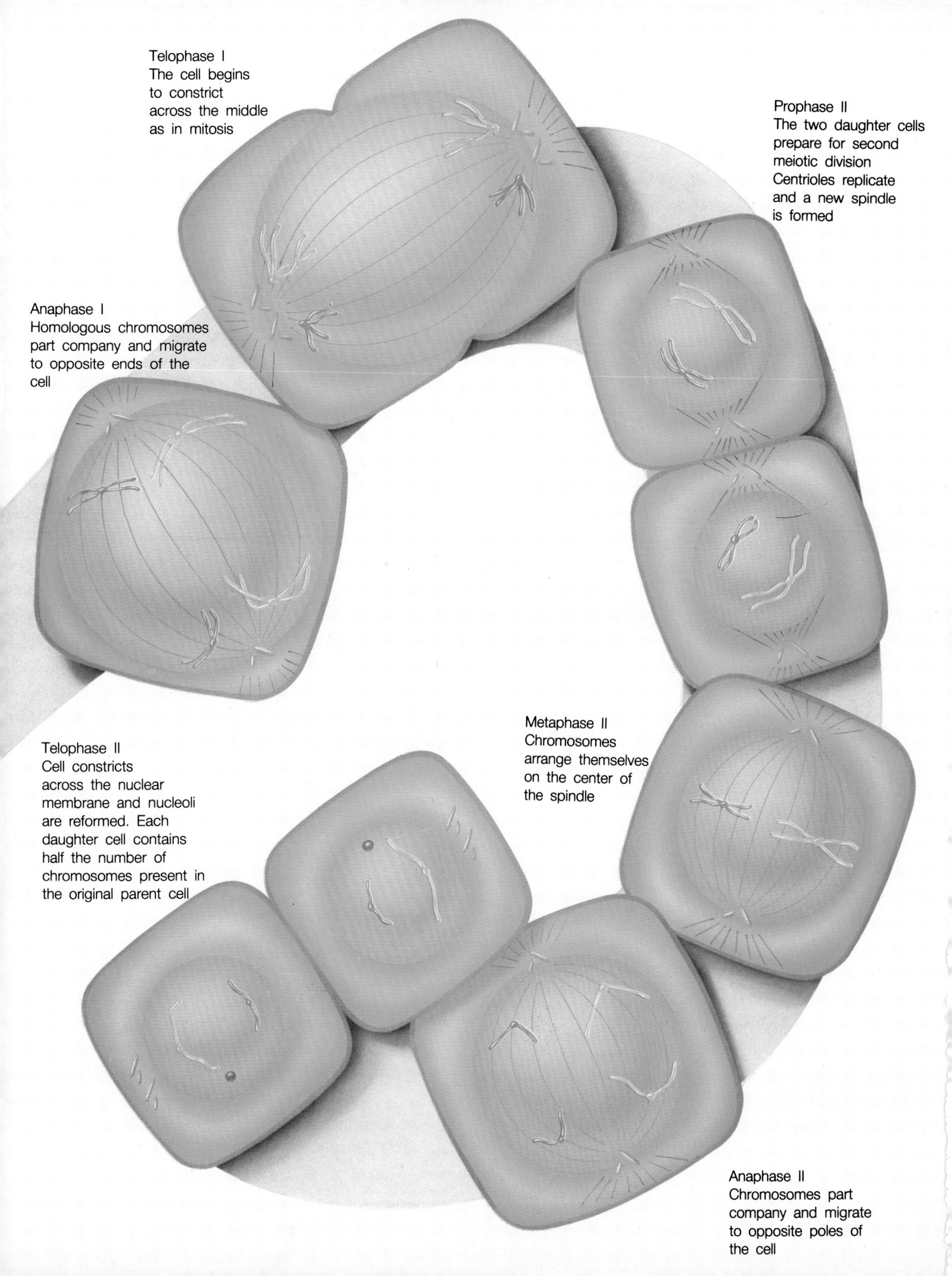
Telophase I
The cell begins to constrict across the middle as in mitosis
Prophase II
The two daughter cells prepare for second meiotic division
Centrioles replicate and a new spindle is formed
Anaphase I
Homologous chromosomes part company and migrate to opposite ends of the cell
Metaphase II
Chromosomes arrange themselves on the center of the spindle
Telophase II
Cell constricts across the nuclear membrane and nucleoli are reformed. Each daughter cell contains half the number of chromosomes present in the original parent cell
Anaphase II
Chromosomes part company and migrate to opposite poles of the cell

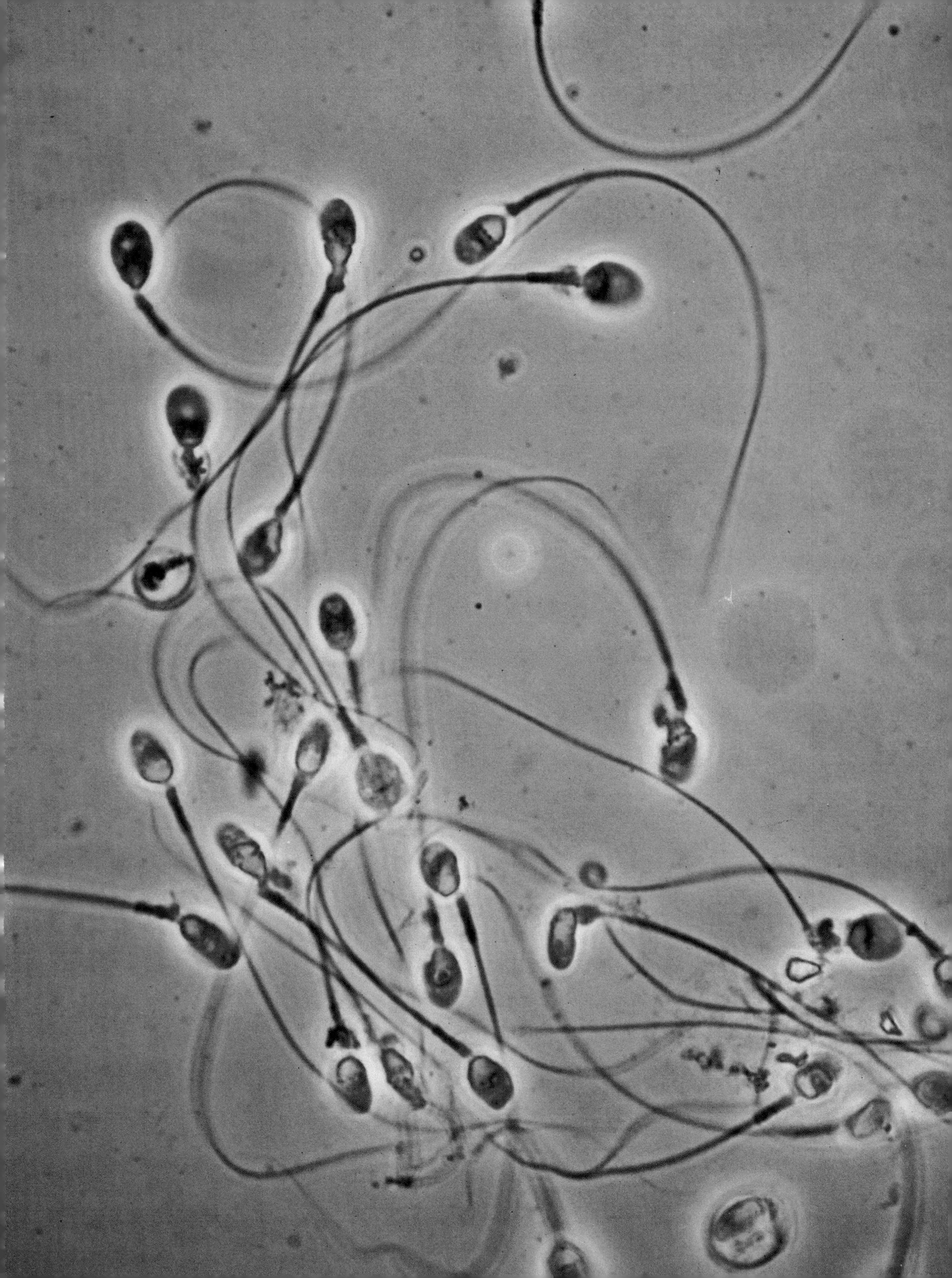

Hermann Joseph Muller

The Master of Mutation

The life and the work of Hermann Joseph Muller both began at a time when Gregor Mendel's research was still not well known, and both ended when virtually all of what is now thought of as basic to the science of genetics had been discovered — much of it as a direct result of Muller's own research and conceptual brilliance. His long life span produced a series of discoveries, of changes in methodology or approach, and of writings that are today commonly regarded as classic.

Muller was born in New York City in December 1890. Upon graduation from Morris High School in the Bronx, he received a scholarship to Columbia University. He finished in 1909 already strongly interested in the mechanics of heredity. As a postgraduate he was then inspired by Thomas Hunt Morgan's introduction to laboratories of the fruit fly as a genetic tool. Even as he was putting together the papers for his doctorate — on the mechanics of the chromosomal behavior known as crossing-over, papers now themselves regarded as very significant — the work of his group was being summarized in a volume entitled *The Mechanics of Mendelian Heredity* (1915), which one commentator has

described as "a cornerstone of classical genetics."

Muller then spent three years at the Rice Institute at Houston. It was apparently here that he first determined to find a way to increase the rate of genetic mutation — for science at this time still relied for data on the slow rate of natural mutations among fruit flies or other experimental creatures. One of his discoveries was that the rate was accelerated by heat, but that both chromosomes of a pair were not necessarily affected identically.

In 1920 he moved to become Associate Professor at the University of Texas at Austin. He was to stay there for a creative twelve years.

It was at Austin that Muller began to experiment with the effects of X rays. He found that the rate of mutation was vastly increased — and with his techniques the whole science of genetics changed gear.

But personal problems were beginning to obtrude. His marriage broke up, his politics were unpopular (he was of strongly socialist views) and his health suffered. So in 1932 he went to work for one year in Berlin, Germany, at the Kaiser Wilhelm Institute, where he sought with others to scientifically determine the nature of a gene.

He was then invited to join the Soviet Institute of Genetics at Leningrad. His socialist views led him to accept, and to move with the Institute to Moscow a little later. But the negative theories of Trofim Lysenko were becoming official dogma in the Soviet Union, and it was only a matter of time before Muller had to leave — which he finally did in 1937.

Edinburgh, Scotland, was the city of his next seat of learning. At the Institute of Animal Genetics there he met and married his second wife. And in 1940 he returned to the United States.

A man truly concerned for the welfare of others, Muller campaigned against the use of radiation of all kinds on humans, until his death in April 1967.

Gout is a chronic disease that produces sudden attacks of swelling in the extremities, particularly the feet. The tendency to gout is hereditary and is not preventable, but can be controlled through diet.

do scientists and counselors base predictions on?

Many of the necessary tools are mathematical and involve complex statistical calculations of probability and frequency. Two types of study can shed direct light on a particular disorder or trait. Family studies can produce a pedigree that traces ancestors and descendants of a particular person for many generations and determines whether they shared a particular disorder. Studies of identical twins allow the comparison of complex gene interactions and the interaction of heredity and environment.

Pedigree analysis is useful for deciding whether or not a particular disorder is likely to be caused by a single gene, and whether that gene is dominant, recessive or sex-linked. These deductions are based on observing that a particular trait is inherited as often and in the same pattern as experimental studies of smaller organisms. It sometimes occurs that a defect is not classically inherited but has "just happened" by mutation. Generally, genetic counselors can nevertheless use pedigree analysis to advise an affected family on their chances of having a damaged child.

This kind of analysis may give an insight into a disorder which spans many years. Huntington's chorea is a type of hereditary brain degeneration transmitted as an autosomal dominant trait. The gene does not cause its deleterious effect until the victim reaches middle age, by which time he or she may already have had children and passed the gene on. In this particular instance, pedigree analysis shows that the gene was brought from Suffolk in England to Boston, Massachusetts, in 1630, possibly by a single servant girl. Some of the hapless victims were accused of witchcraft in the Salem witch trials a generation or two later.

The Mystery of Penetrance

Twin studies are a fascinating part of heredity because identical (monozygous) twins have identical genes. They result when a single fertilized egg divides into two separate embryos early in intrauterine life. (Fraternal or dizygous twins occur when two eggs are released and fertilized instead of the usual single egg, and these twins are no more genetically alike than ordinary brothers and sisters.) Identical twins have the same birthdate and ordinarily the same environment, which means that any subtle but differing effects of heredity and environment can be studied. Many polygenic traits can be studied accurately only in this way. Certain illnesses much more commonly affect both identical twins than both fraternal twins. If one identical twin suffers from schizophrenia, for example, then there is a 60 to 80 per cent chance that the other twin also has the disorder. The same situation with a non-identical twin gives the other twin only a 10 per cent chance of having the illness — strong evidence toward an important genetic component in this disorder.

However, this same study reveals some of the complexity surrounding genetics. Despite identical genes and a clear genetic component, not all cotwins of schizophrenics develop the disorder. No one knows why some people are protected. Perhaps not all the genes they inherit fully express themselves. This phenomenon in which genes only partially show their effects is called incomplete penetrance. It complicates genetic predictions

In certain isolated groups like the Amish people (below), *there is a high rate of intermarriage. Marriage to near relatives may increase the number of offspring with physical or mental disorders* (right).

because it means that genes do not necessarily have the same effect in different people, and for the moment remains a mystery.

Gout is one such condition. It is known to run in families and be related to the blood concentration of uric acid, a metabolic breakdown product. An affected person has a high uric level all the time, but only occasionally gets attacks of painful joints. Some people have similar high uric acid levels but never get gout attacks, even when their relatives are subject to them. Thus gout, which may be caused by a single dominant gene, is said to show variable expression or incomplete penetrance. Not everyone who has the faulty gene has the condition.

Human heredity is a fascinating, complex and largely abstruse field of knowledge. The effects of relatively few single genes are clearly understood, but little is known about the interactions between multiple genes. The physical features of our future offspring are unpredictable, and there is little one can do to influence them. So far even trying to choose the sex of a child has proved impossible.

Chapter 3

Cracking the Code

Occasionally the entire scientific community goes off on completely the wrong tack — and this is exactly what happened from the 1920s up to 1953 in connection with the substance considered to be the chemical basis of heredity. Up to that time scientists thought that proteins were the hereditary substance. Proteins were complex and numerous, whereas DNA was so simple that it was boring by comparison. This threadlike molecule with a tendency to form sticky lumps and composed of only four subunits could surely never contain all the diverse information necessary to build even a simple cell. And as for building a complete animal — this was thought to be far beyond the capacity of such a simple structure. Only proteins seemed capable of containing the "library" of inheritance.

It now seems odd that such wrong beliefs could arise, for the actual experiments leading to the discovery of the chemical basis of heredity had by then already been performed. Unfortunately, though, the results had been largely ignored.

The inheritance of obvious facial similarities — these are members of the Bridges family, painted by John Constable — prompted geneticists to study other inherited characteristics. Among the first to be investigated were anomalies such as polydactyly (extra fingers), albinism and color blindness. It is now known that polydactyly is caused by a single dominant gene, albinism involves a combination of two recessive genes, and color blindness — like hemophilia — is an example of a sex-linked inherited trait. The next step was to determine the chemical basis of heredity: to find out what genes were made of.

Crucial Experiments

In the 1940s, pneumococcal pneumonia (now called by the name of its infective organism, *Streptococcus pneumoniae*) was a leading cause of death in children. One form of treatment in the era before antibiotics was to administer to an affected individual a copious dose of antibodies (made in horses) to the exact strain of bacteria causing the infection. It was thus essential that the strain of pneumococcus was correctly identified, or the wrong antibody could be given. The possibility that a bacterium might change its coat and escape, or give misleading results when identified, was a cause of major concern. Great interest was generated, then, when experiments showed that capsular change (the bacterium's changing its coat) could occur when DNA from one strain of bacteria was added to a totally different strain.

The second crucial experiment, in 1952, proved

Pneumonia bacteria gave some of the first clues that DNA was the substance of heredity when scientists noticed that the bacterium "changed its coat" when DNA from one strain was added to another.

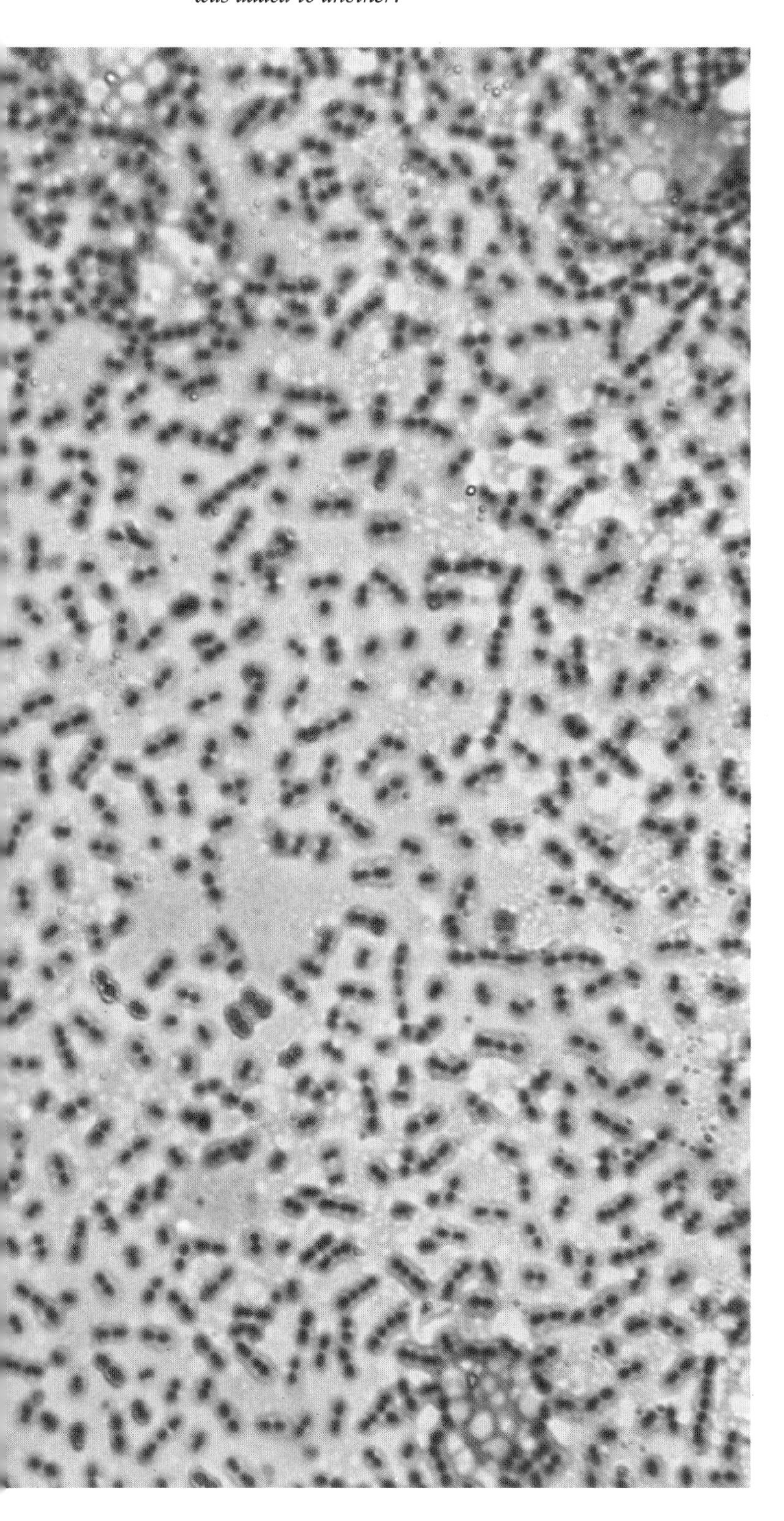

that DNA could be the "hereditary principle." Scientists discovered very simple viruses called phages, which infect bacteria and which could be grown. In a now-classic experiment, the American biologists A. D. Hershey and Martha Chase labeled these viruses with radioactive phosphorus or radioactive sulfur. This was ingenious because DNA contains large amounts of phosphorus but no sulfur at all. Proteins are just the reverse — some contain sulfur but none contains phosphorus. It soon became clear that the phosphorus-containing molecules were passed onto progeny viruses but the sulfur-containing molecules were not. In this simple organism, DNA must be the hereditary factor; protein was not transmitting to offspring.

When James Watson and Francis Crick made their dramatic statement about the structure of DNA, and how it could serve as the molecule of heredity, they totally transformed scientific thinking. In a single paper, which contained no experimental data, they both predicted the exact chemical structure of DNA and showed how the genetic information was stored. They did not break the genetic code — that came later — but they showed that the code existed, which "library" held it, and how it was written. It is possible that in this century the two discoveries that will determine humankind's fate are the discovery of the power of the atom and the discovery of the genetic code.

Before this breakthrough, genetics was dominated by experiments concerning the inheritance of gross anatomical features like eye color or wing shape in fruit flies. Afterward the inheritance of simple metabolic events, such as the ability to grow under special conditions, dominated the scene. The reason is that, in the 1940s, Joshua Lederberg and Edward Tatum had discovered that bacteria have a kind of sexual reproduction — instead of always dividing into two identical progeny, bacteria sometimes pass genetic information from one cell to another. Because this exchange occurs in an orderly, predictable way, an elegantly simple experiment was eventually designed which allowed bacterial genetics to become a much better tool for studying heredity than flies or mice, which until then had constituted the major subjects for genetic experimentation over a period of approximately four decades.

When a cell divides, the chromosomes in the nucleus become tightly spiraled and rodlike (below left). *As division progresses they are drawn to the center of the nucleus* (below right).

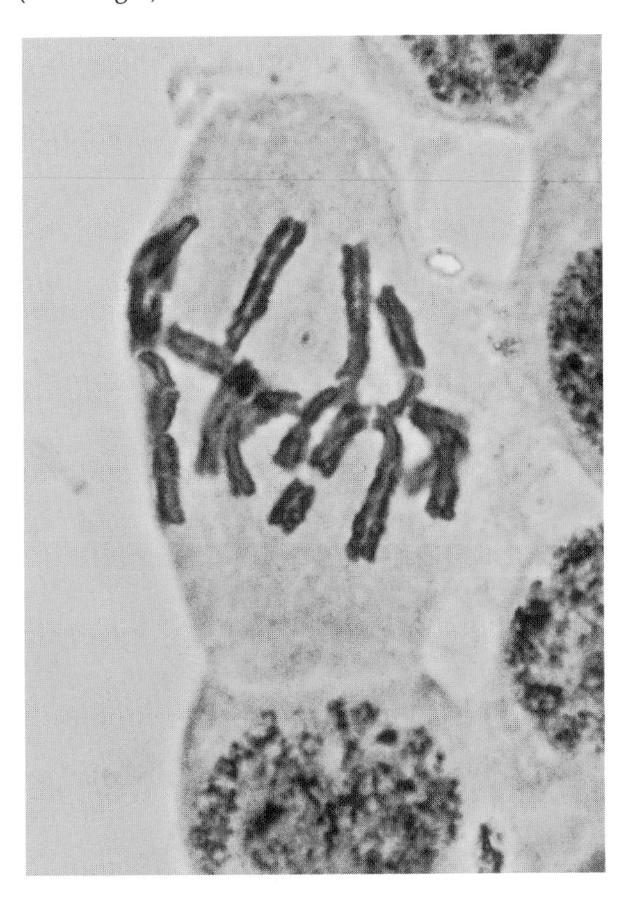

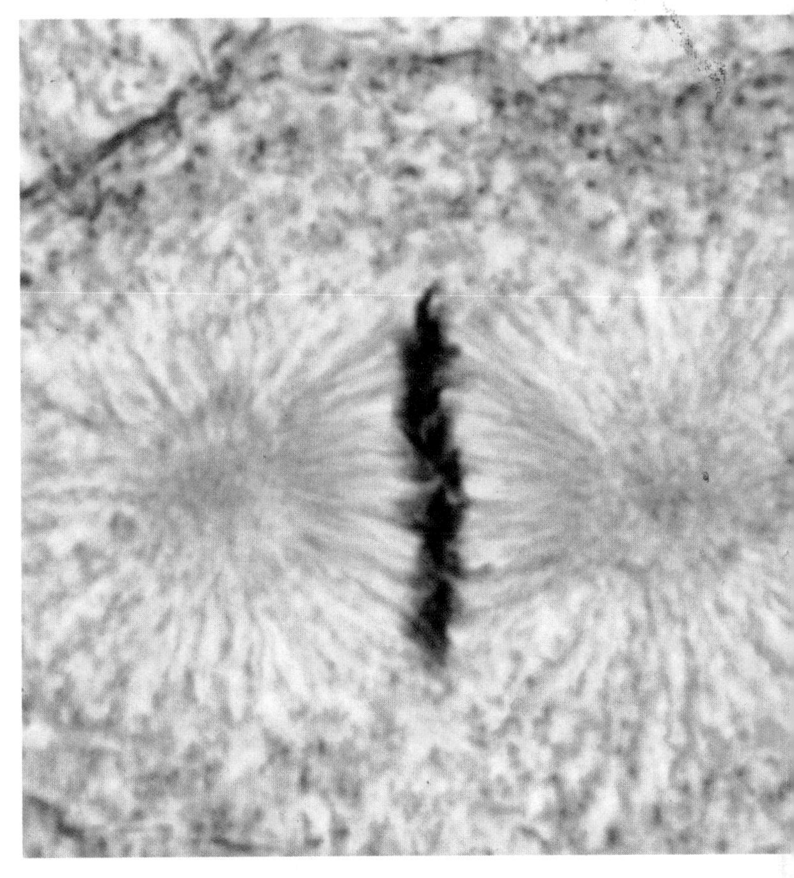

The simple trick was to mix the bacteria in a blender at a specific time after this process of reproduction (conjugation) had started. Mating pairs were disrupted and the effect on the progeny measured. What emerged was an orderly transmission of information from one cell to another, the same gene (or characteristic) was always transferred first and the same one last. Bacteria appeared to have linear genetic material arranged in a definite order. Following this discovery of a bacterial chromosome, and together with the realization that all living organisms used DNA for their hereditary "library," it became possible to study genes and their effects in minute detail.

To properly appreciate the power of the new genetic analysis, it is helpful to examine the situation up to 1953. At the turn of the century the close similarity between the behavior of chromosomes and the inheritance of Mendel's "factors" was noted. Chromosomes appear like rods or drumsticks in the nucleus of a cell when it divides. Having duplicated before cell division, they are pulled apart during the division so that each new cell gets exactly one complete set. This process of mitosis is, in most cells, the only time the chromosomes are visible. Usually they are invisible, and no one is certain how they are arranged when the cell is not dividing.

There is one exception to the normal process of mitosis; this occurs in connection with division of the sex cells or gametes, a male and a female of which fuse to produce offspring. Such cells must contain one-half of the normal complement of chromosomes. This is accomplished by splitting the chromosomes two times: one to separate the duplicate chromosomes into separate cells, and a second to separate the members of a chromosome pair into daughter cells. This process of meiosis is fundamental to understanding how genes get reassorted during sexual reproduction.

One of Indira Gandhi's most distinguishing features was her white forelock. The gene for this trait is thought to interfere with the formation of pigments in scalp cells, leading to the growth of white hair.

During meiosis each pair of chromosomes separates by chance into the new sex cell. For example, if the chromosomes pairs could be labeled, Aa, Bb and so on, the pair Aa would separate into A and a and the pair Bb into B and b. Whether A or a goes into a particular sex cell is a matter of chance, and the same is true for B and b. So there could be A and B, or A and b, or a and B, or a and b occurring together in the newly-formed egg or sperm. There are twenty-three pairs of chromosomes in humans so that there are 2^{23} or more than ten million different ways the chromosomes can be assorted during meiosis.

When an egg and sperm meet, the resulting fertilized egg has two chromosomes in each pair, one from each parent. The chances of two children of the same parents getting exactly the same set of each is so small that it effectively never happens. Only identical twins, who start off as a single fertilized egg, can have identical chromosomes. Because chromosomes behave like genes scientists decided that genes were probably attached to the chromosome.

One prediction of this chromosome theory was that some traits should always occur together because the genes responsible for them should be linked together on the same chromosome. Sure enough, linkage of some traits—but not others—was found. Again genes behaved as if they were on chromosomes, and different chromosomes carried different groups of genes. This was an important idea; if the chromosome theory was correct it should have been possible to make linkage maps, showing groups of genes that traveled together, and even to find out which chromosomes contained specific gene combinations.

Fruit Flies Give Further Clues

Fruit flies had for years played the central role as the geneticists' most useful tool in the development of the chromosome theory. But there was a major snag: how to find out which genes belonged to each chromosome.

One useful property of chromosomes is that they bind certain dyes in a very regular pattern of bands. Each pair of chromosomes has its own unique banding pattern, and so can be distinguished from all other pairs. Occasionally a mutation, particularly one induced in the laboratory by radiation, caused a small but distinct change in the tell-tale pattern on one chromosome. This was like labeling that chromosome with a specific marker and it gave the scientists a handle on that chromosome.

Now they could follow it around in breeding experiments and see which inherited characteristics went with the change. Soon whole maps of linked genes were made for all the fruit flies' chromosomes. A model of heredity emerged from these experiments. Genes were perceived to be like beads or pearls in a necklace, and mutations came along and either altered a bead like changing the color and value of a pool ball amidst other pool balls, or knocked a bead off the string like a pool ball being knocked into a pocket.

The ability to band chromosomes with dye has revolutionized the study of those in humans. Although the twenty-three pairs can be subdivided into about seven groups with similar sizes, it is very

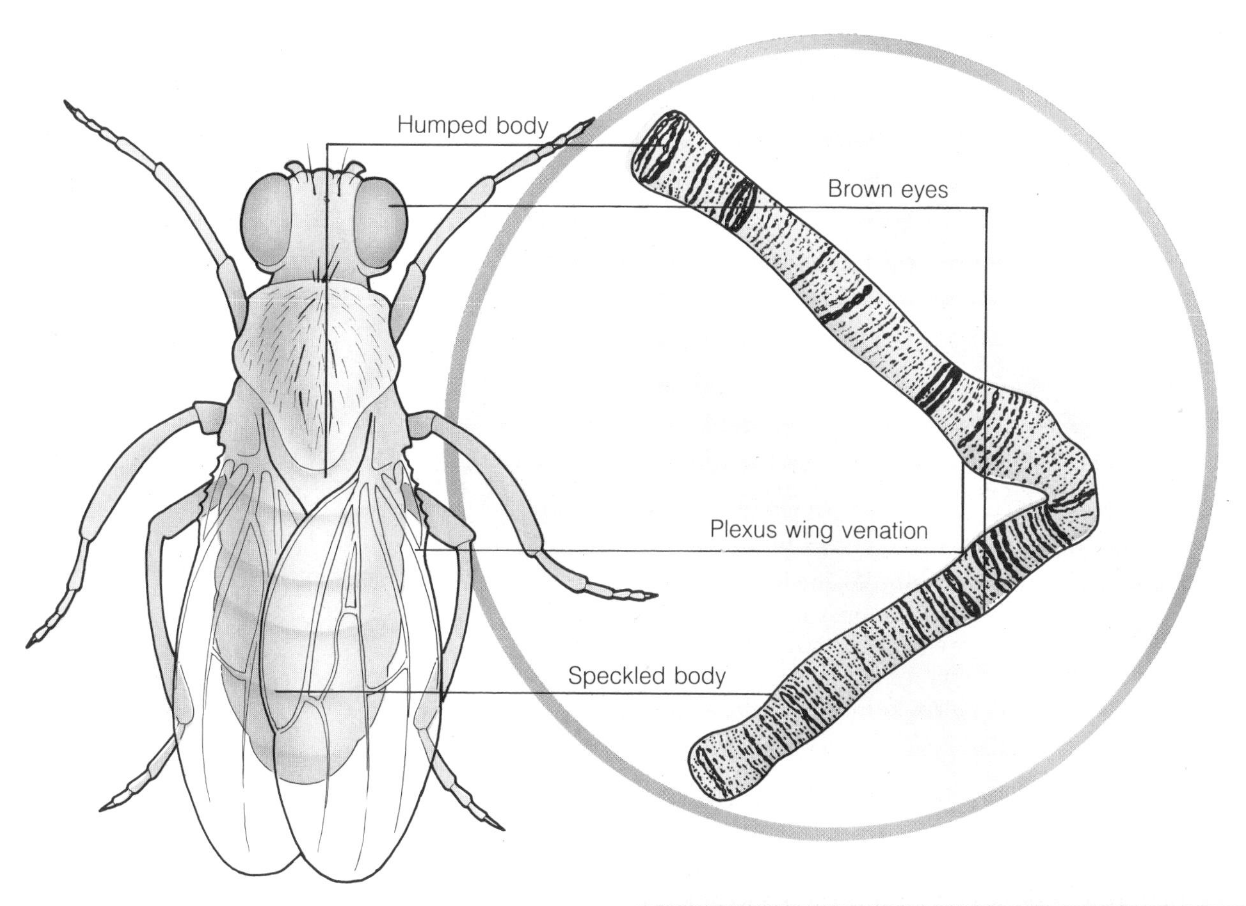

The giant salivary gland chromosome of the fruit fly Drosophila melanogaster (above) *has allowed scientists to come closer to locating genes on a chromosome. The DNA occurs in tightly coiled bands along the chromosome. Dyeing techniques applied to the DNA have revealed the approximate location of the genes responsible for features such as a humped body, a speckled body, plexus venation and brown eyes. Scientists have been able to study mutations in fruit flies, for example white eyes as opposed to the normal red* (right). *Because of the rapid breeding ability and short life span of these insects, several generations can be produced and observed in a relatively short period.*

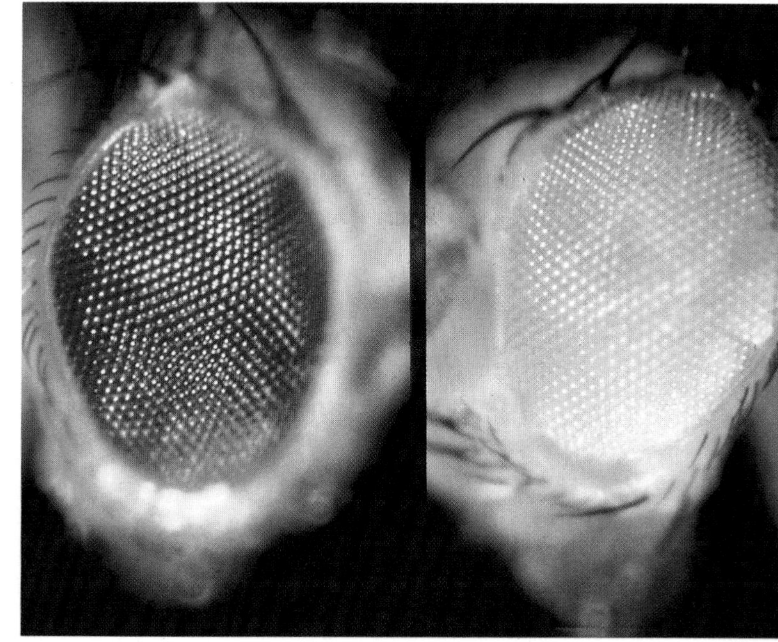

difficult to tell apart the members of a group which are very similar in length. Banding identifies each chromosome precisely so that even those which otherwise appear identical can be distinguished. Only by this method was it possible to agree on numbers for each chromosome and to identify the same chromosome in different individuals. Once it was possible to identify each normal structure, it became clear to scientists that abnormal chromosomes sometimes did occur and were associated with many different disorders.

In general chromosomes have similar structures but with variations on a theme. The DNA and protein is packaged into arms. There is a narrow area or constriction on each chromosome called the centromere. This can be anywhere along the length of the chromosome so that each one has an arm above and below the centromere. The two arms can generally be identified as the short arm and the long arm and, along with the banding pattern, help to identify the subregions of each chromosome. Scientists have also given numbers to the bands found on each arm so that they can identify abnormal chromosomes very precisely.

So far the chromosome story may seem simple but, as with all things in nature, accidents can happen: the mechanism of dividing chromosome pairs during meiosis and mitosis can go wrong. In humans this can have catastrophic effects. Sometimes the chromosome pairs do not separate equally; instead of one going into each new cell during division they may get stuck together. Two go into one daughter cell and the other cell misses out. In ordinary body cells (somatic cells) this is not too disastrous because usually both abnormal cells die and are replaced by their normal cousins. But in a sex cell real problems can occur, especially if the accident happens in an egg cell.

Chromosome Abnormalities

Egg cells are formed early in the life of a female child, and once formed they are not replaced — although they mature only just before being shed at ovulation. By contrast, sperm cells are rapidly lost and replenished. Age increases the chance of slightly damaging a maturing egg so that at the completion of meiosis the two members of a chromosome pair do not separate fully, leaving cells with an excess or deficiency of chromosomes. Most cases of nondisjunction are not viable, and an affected embryo is rapidly and naturally aborted. About half of all spontaneous abortions (miscarriages) are probably due to the embryo's having either too many or too few chromosomes. The increased chance of this in older women accounts in part for the decreased fertility of women over the age of thirty.

Genes are located at specific sites along the chromosomes. These fruit fly chromosomes have been dyed to make the gene locations — visible as light and dark bands — easier to study using a microscope.

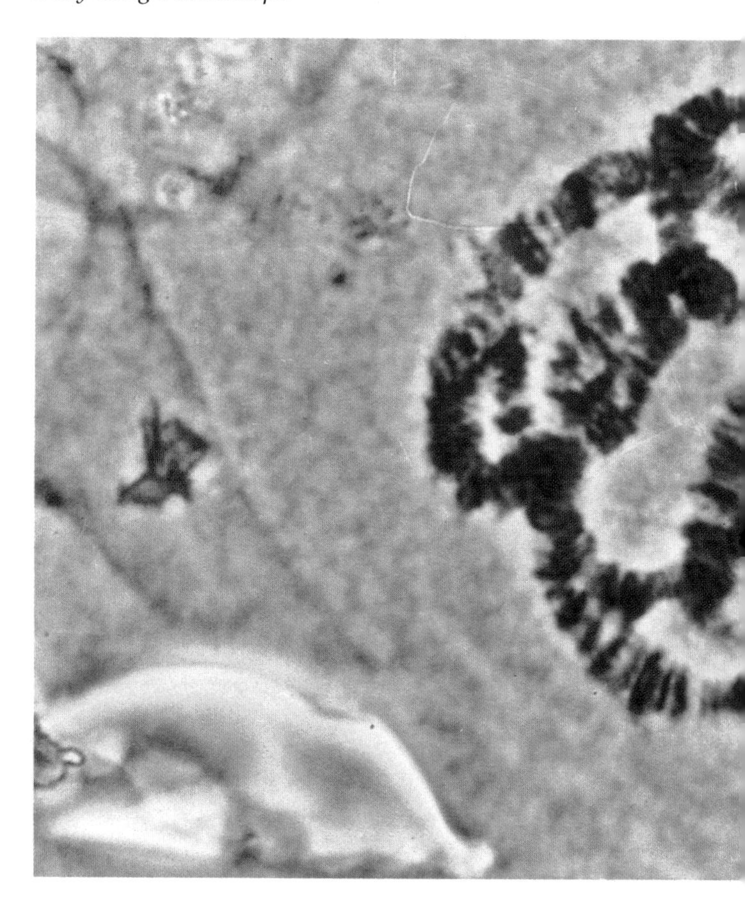

One particular example of nondisjunction is often compatible with life, however; the fetus survives, and a child is born. What happens is that during the formation of the egg, both number 21 chromosomes end up in one egg cell which, when fertilized by a sperm carrying its single chromosome 21, produces a child with three 21 chromosomes per cell. Having three of a chromosome is known as trisomy, and trisomy 21 results in Down's syndrome, sometimes called mongolism. Affected children are usually short in

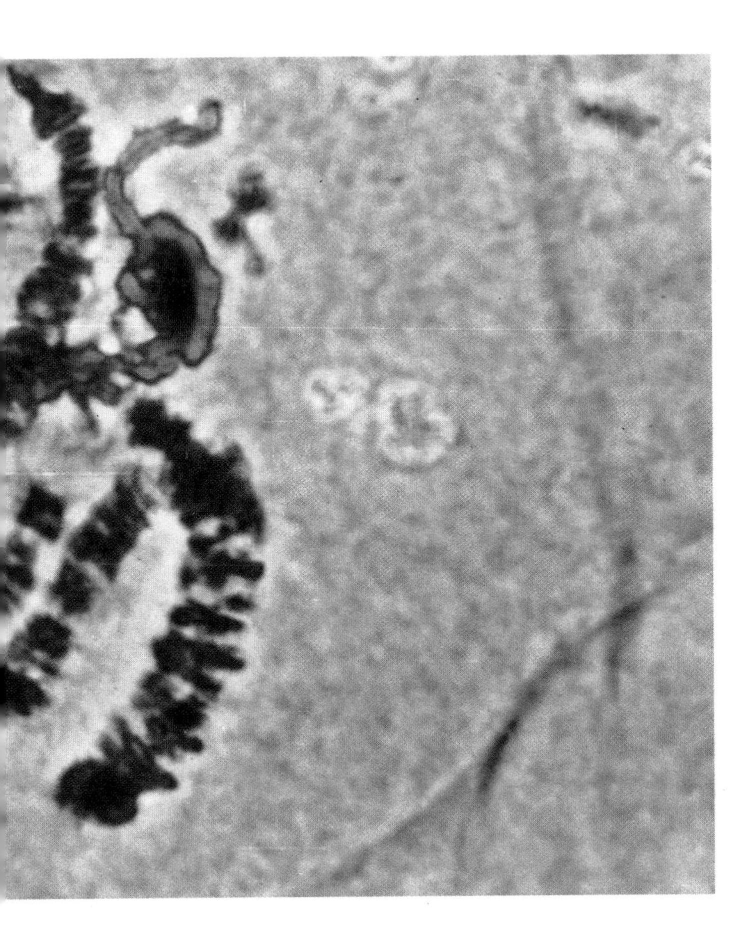

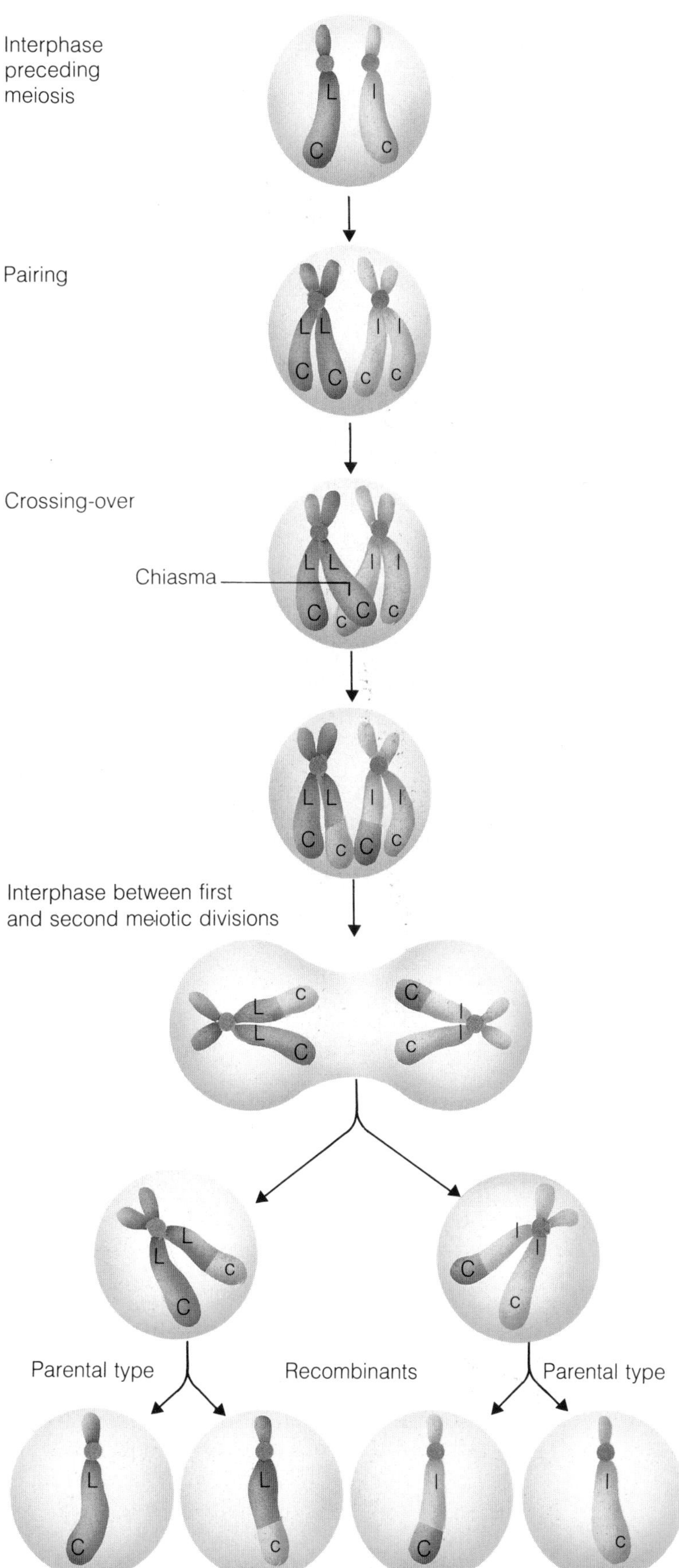

Early on in cell division chromosomes pair up and two pairs cross over at a point called the chiasma. In this case the dominant allelomorphic gene, or allele, denoted by C, for curly hair, crosses with the recessive allele c, for straight hair (the alleles L and l are for a long and short nose respectively). Alleles C and c change over onto adjacent chromosomes which separate again and the cell divides. Following a second division the chromosomes split up so that they are single (haploid) once more, but two of them differ from the parental types L/C and l/c. Thus crossing-over, which affects all the many thousands of genes on a chromosome, creates genetic variability.

stature, have an extra fold of tissue on their eyelid next to their nose, and are mentally retarded.

Down's syndrome is probably the most common congenital abnormality, and affects about one in every six hundred children born alive. The chances of having a child with trisomy 21 increases with the mother's age, so that about one per cent of children born to mothers over the age of forty are affected. This has an important influence on the advice given by genetic counselors and the types of tests that might be carried out on a pregnant woman at different ages. Although the relationship between maternal age and Down's syndrome babies was known in the nineteenth century, the exact chromosomal abnormality was found only in 1959.

Apart from trisomy 21, excesses or deficiencies of whole chromosomes are uncommon, and only numerical abnormalities of the sex chromosomes X and Y seem compatible with survival for more than a few weeks. These are also fairly common and often less devastating than other chromosomal problems. A male with the combination XXY, called Klinefelter's syndrome, is usually lanky, mentally retarded and sterile. Another combination, XYY, also produces tall men but who are fertile, and who at one time were suspected — possibly quite unfairly — of being particularly likely to be of a violent or criminal disposition because there seemed to be more men with this makeup than expected in jails. Approximately one child in a thousand has some type of sex chromosome abnormality. Often it is not detected until the person experiences difficulty in having children.

Trisomies involving other chromosomes occur, but only rarely in non-stillborn children. Even then these children have such severe abnormalities that they rarely survive for more than a few weeks.

Children affected by trisomy and related numerical abnormalities of chromosomes cause some of the most complex and heartrending problems of

A woman who becomes pregnant after the age of thirty-five (left) *has a greater risk of giving birth to a child with trisomy 21, or Down's syndrome* (below), *than have women of a younger age.*

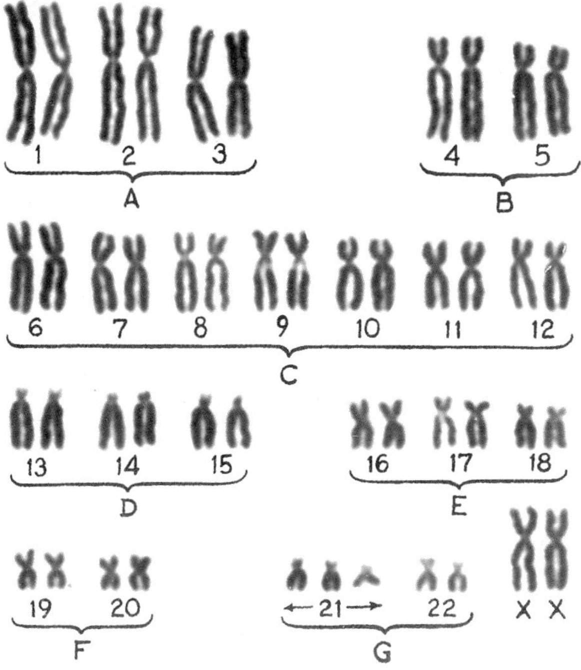

Trisomy 21 is caused by a chromosomal anomaly, as revealed by the chromosome map (left). *This shows that the chromosome pair 21 (in the G set) has an extra chromosome. It occurs because a pair of chromosomes in the egg, which would normally separate, remain together and are then joined by the chromosome from the sperm cell at fertilization.*

Chromosomal abnormalities can lead to disorders such as a type of leukemia. This results from a chromosomal exchange that creates an imbalance of genetic material. Affected cells are dark pink.

Abnormal numbers of chromosomes can cause personality disorders. For example, men with the XYY combination were thought to be violent, as depicted by this actor in a television series about an XYY man.

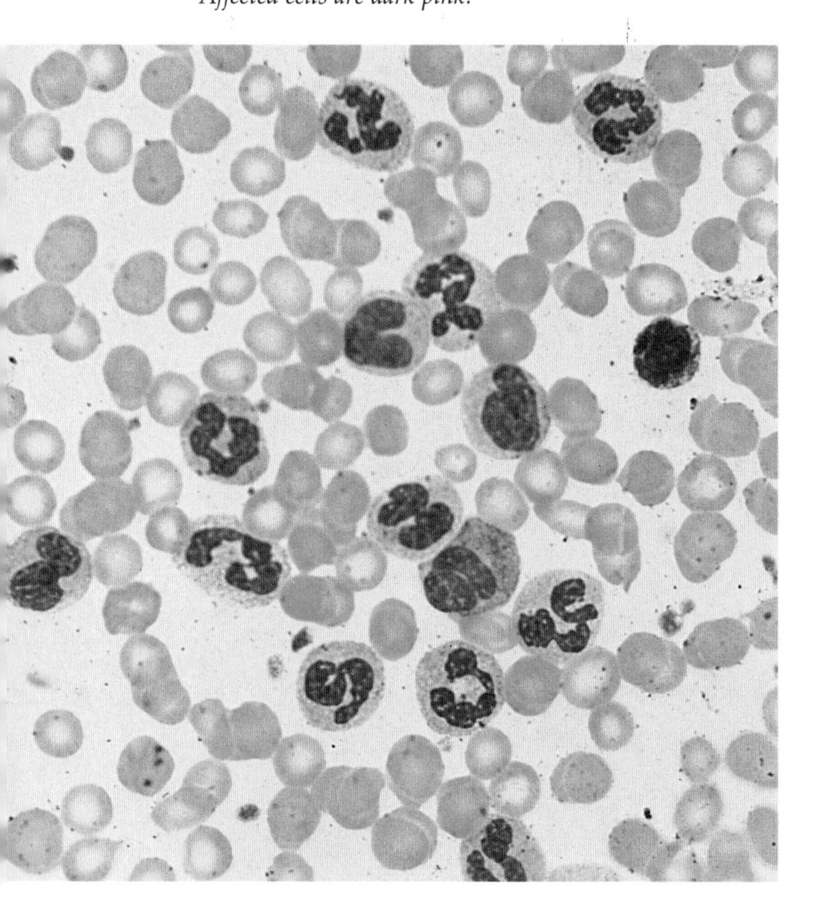

modern medicine. Down's syndrome children often have further physical abnormalities, such as abnormal hearts or blocked intestines. Should they be repaired so as to allow the child to live an abnormal life, or is it a greater kindness not to intervene and to let nature take its course? Who should make this kind of decision, the parents, the doctors or the courts? If parents of a child with this type of genetic abnormality honestly feel that the child should be allowed to die, who has more right then they to decide differently?

Nature sometimes makes a mistake when packaging chromosomes and building new copies of the old models. Instead of too many or too few, this results in chromosomes which are rearranged. In the language of genetics, translocations occur when two chromosomes of different pairs exchange segments. If the exchange is complete and the amount of genetic material remains normal, the translocation is said to be balanced. However, the person carrying the translocation may pass on one of the rearranged chromosomes to his or her offspring, which produces an unbalanced state equivalent to an excess or deficiency of a partial chromosome. Inversions occur when a portion of a chromosome gets rearranged within the same chromosome. Often this sort of abnormality may go unnoticed. But sometimes a translocation of chromosomes is associated with the development of a cancer.

One such translocation affects chromosome 22. Part of chromosome 22 gets exchanged with a small piece of chromosome 9 so that the one chromosome is shorter than its normal counterpart. This abnormality is frequently found in one kind of leukemia called chronic myeloid leukemia. The damaged chromosome condition can be inherited, but more commonly arises by chance in somatic tissue.

This close association between one type of chromosome damage and one unusual type of cancer led to a search for others. It appears that rearrangement of a particular group of genes often produces abnormal growth of that kind of cell. Identifying such markers can be used to follow the effects of treatment. If the genetically abnormal cells associated with chronic myeloid leukemia disappear during treatment, then the leukemia has been successfully — if only temporarily — treated, and if they reappear it is an indication for further treatment.

Some cancer-associated translocations seem to relocate members of a particular family of genes called oncogenes into abnormal places, with the result that the normal genetic "control switches" for growth are altered.

Human gene maps detailing which genes are on specific chromosomes have been hard to make because until recently there was no way to localize any one gene to a particular chromosome. Occasionally a family study would show that two genes were linked, so that if one gene could be localized the other was also known to be on the same chromosome. This was a very tedious method and located only a few genes precisely. However, a new "trick" of cell biology has produced an explosion of knowledge.

This trick involves the Sendai virus — a distant

cousin of the measles virus — or polyethylene glycol, both of which have the useful property of making cells join together into one hybrid cell containing all the chromosomes from both parents. When cells from humans are fused with mouse cells, the chromosomes get reshuffled like a deck of cards and many of the human chromosomes are thrown out.

This may seem a meaningless kind of experiment, but the result is extremely useful because in the end cells with mostly mouse and only one or a few human chromosomes are left. A set of hybrid cell lines each containing at least one different human chromosome has been built up, so that all twenty-three is represented in at least one cell. Now each chromosome can be studied individually. With the use of gene probes (labeled pieces of specific genes, as explained in the next chapter), it is possible to locate genes exactly. This ability has revolutionized the study of how genes travel together and of where specific genes, such as those implicated in cancer, are located.

The Properties of Genes

Chromosomes are composed of long strands of DNA with many proteins attached. When Watson and Crick showed that DNA and not protein was the hereditary substance, attention was turned to bacterial chromosomes to study how this structure functioned. The chromosomes in these organisms are often simple circles of DNA that can be seen using an electron microscope.

The chromosomes of animals and plants are, by comparison, extremely complex. The DNA is tightly coiled around itself like a rubber band used to power model airplanes. The tightly wound molecule is then wrapped around proteins like thread around a wooden spool, and finally several spools are packed together as if in a box or mailing tube. Unlike spools of thread in a box, however, all

Some fungi, such as the one shown here growing on human skin, need the amino acid arginine in order to grow. Arginine was used in early genetic studies of the relationship of amino acids with cell growth.

Uncoiled DNA resembles tangled spaghetti in this false-color transmission electron micrograph. This contrasts with its tightly coiled nature when it forms chromosomes, making them more rodlike in shape.

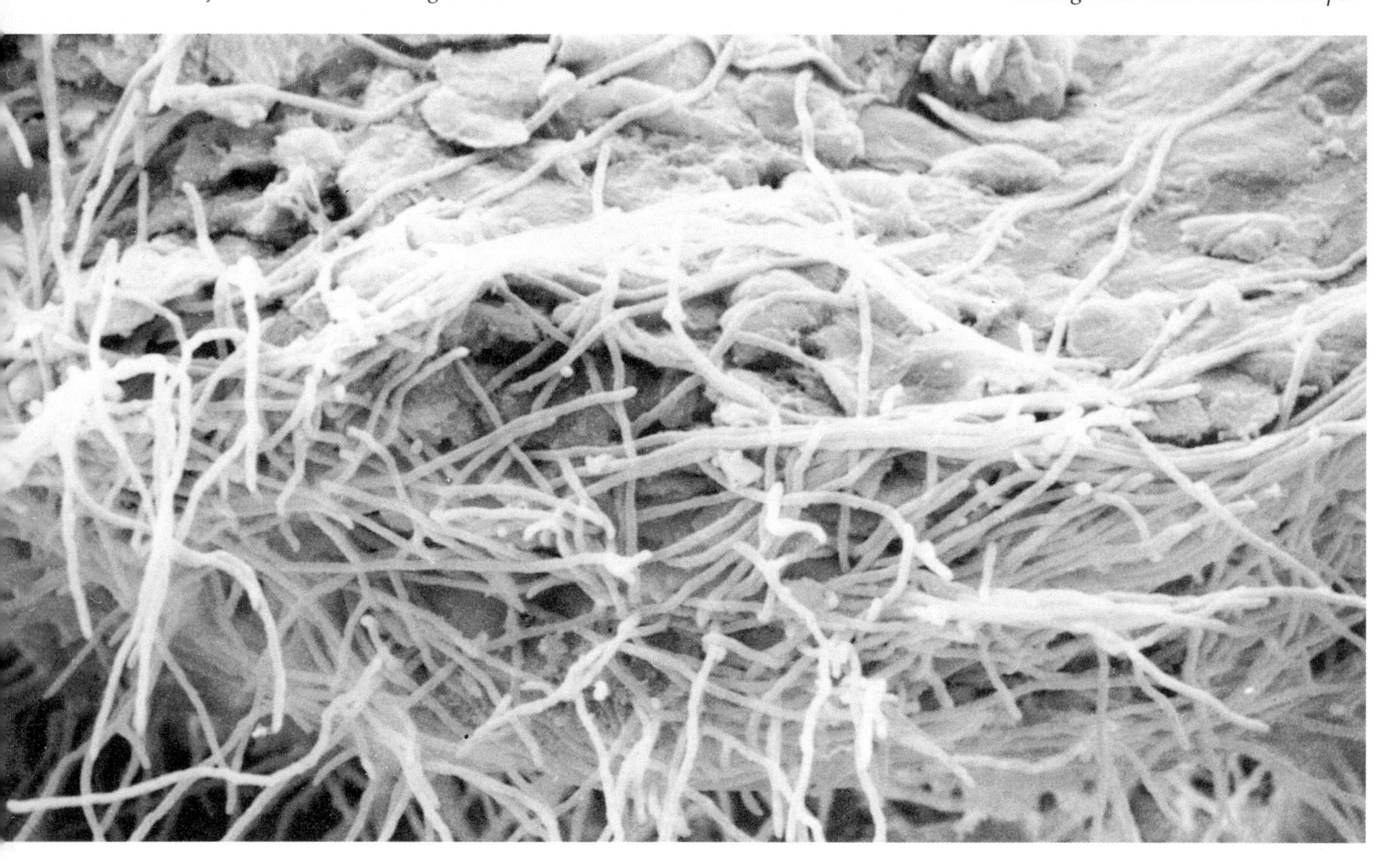

the DNA is a continuous strand as if one enormously long thread were wrapped around many spools.

The simple chromosomes of bacteria have shown how DNA is replicated (reproduced) and how genes are attached on the string. They have made possible the analysis of what a gene is and have allowed the genetic code to be cracked. All of these discoveries have direct effects on medicine. They are also fundamental to the development and growth of the biotechnology industry.

Perhaps the biggest surprise was the discovery that genes were not just like beads on a necklace. They had their own internal structure, and pieces of a gene could be shuffled around to make new and interesting combinations. Another discovery was that mutations did not always have the same effect on a gene: there were many different ways that a gene could be damaged, some much more serious than others. This gave clues as to why two individuals with abnormalities of the same gene may suffer from very different disabilities or disorders, or sometimes not be affected at all. Their mutations affect different parts of the same structure.

Microbial genetics allowed scientists to find out exactly what a gene did. It is now possible, for example, to select fungi that require the amino acid arginine in their nutrient supply in order to grow. The two scientists George Beadle and Edward Tatum collected several different arginine-requiring fungi, and showed that their mutations could be divided into three groups depending on which other amino acids they could use besides arginine. These three groups mapped to different places in the fungal chromosome, indicating three different genes. The compounds each type used to make arginine could be arranged in a kind of sequence because they formed a metabolic chain —*A* is converted into *B* and then *B* is converted into

Edward Tatum

Founder of Biochemical Genetics

Rarely has one researcher contributed so much to the establishing of a new scientific field as has Edward Lawrie Tatum. Indeed, his significance to the progress of biochemical genetics is dual: with one collaborator he prepared the basic concepts of the science, and with another he then provided information and a methodology that has been in use ever since.

Tatum was born in Boulder, Colorado, in December 1909. Because his father was Professor of Pharmacology at the University of Wisconsin, it was there that he studied chemistry, obtaining his master's degree (in microbiology) in 1932. Five years later he was appointed research associate at Stanford University, California.

At Stanford he came into contact with another American geneticist, George Beadle. With Beadle, Tatum successfully established the importance of genes in all biochemical processes involving sequences of chemical reactions. In the pink bread mold, *Neurospora crassa*, Tatum and Beadle induced mutations through the use of X rays. These mutations resulted in specific nutritional deficiencies which the two men demonstrated to be caused by a lack of particular enzymes normally present to catalyze

the appropriate reactions. Different mutations produced different types of deficiency. The inference was obvious: each gene is somehow responsible for one specific enzyme, and thus contributes to the overall regulation of chemical processes.

This discovery was of topical value: the genetic regulation of certain chemical processes considerably facilitated maximizing the production of penicillin during World War II.

At the end of the war, Tatum moved to Yale University. There he continued to investigate nutritional mutations—not, this time, on bread mold but on yeasts and bacteria. He also worked in collaboration with a new partner, Joshua Lederberg, at that time only a student at Columbia University Medical School. Within the space of one year, Tatum and Lederberg were able to demonstrate that bacteria had the power to reproduce by combinative generation, almost sexually.

Because bacteria are far easier to work with than, for instance, fruit flies, and the interval between "generations" is so much shorter, bacteria are now standard subjects for genetic investigations.

In 1948 Tatum returned to Stanford University. Lederberg did not return to his student classes, but instead took up senior posts elsewhere, and eventually succeeded Tatum at Stanford. In 1957 Tatum was offered a position at the Rockefeller Institute for Medical Research (now Rockefeller University), in New York City, which he accepted.

His pioneering and formative work was finally given international recognition through the award of the Nobel Prize for physiology or medicine in 1958, which he received together with George Beadle and Joshua Lederberg.

He remains closely associated with Rockefeller University.

The one gene-one enzyme experiment involved exposing a bread mold to X rays to mutate its offspring. A spore from the mold was placed in a medium complete with nutrients and developed into a mold. Spores from this new mold, which were put in a medium with few nutrients, did not grow, and those put into a minimal medium with added amino acids did grow. This proved that the mutant strain needed extra amino acids for growth, unlike the normal mold. A spore from the third mold was then placed in minimal mediums, each with a different amino acid, and developed only in E. Thus E was the amino acid the mold could not make.

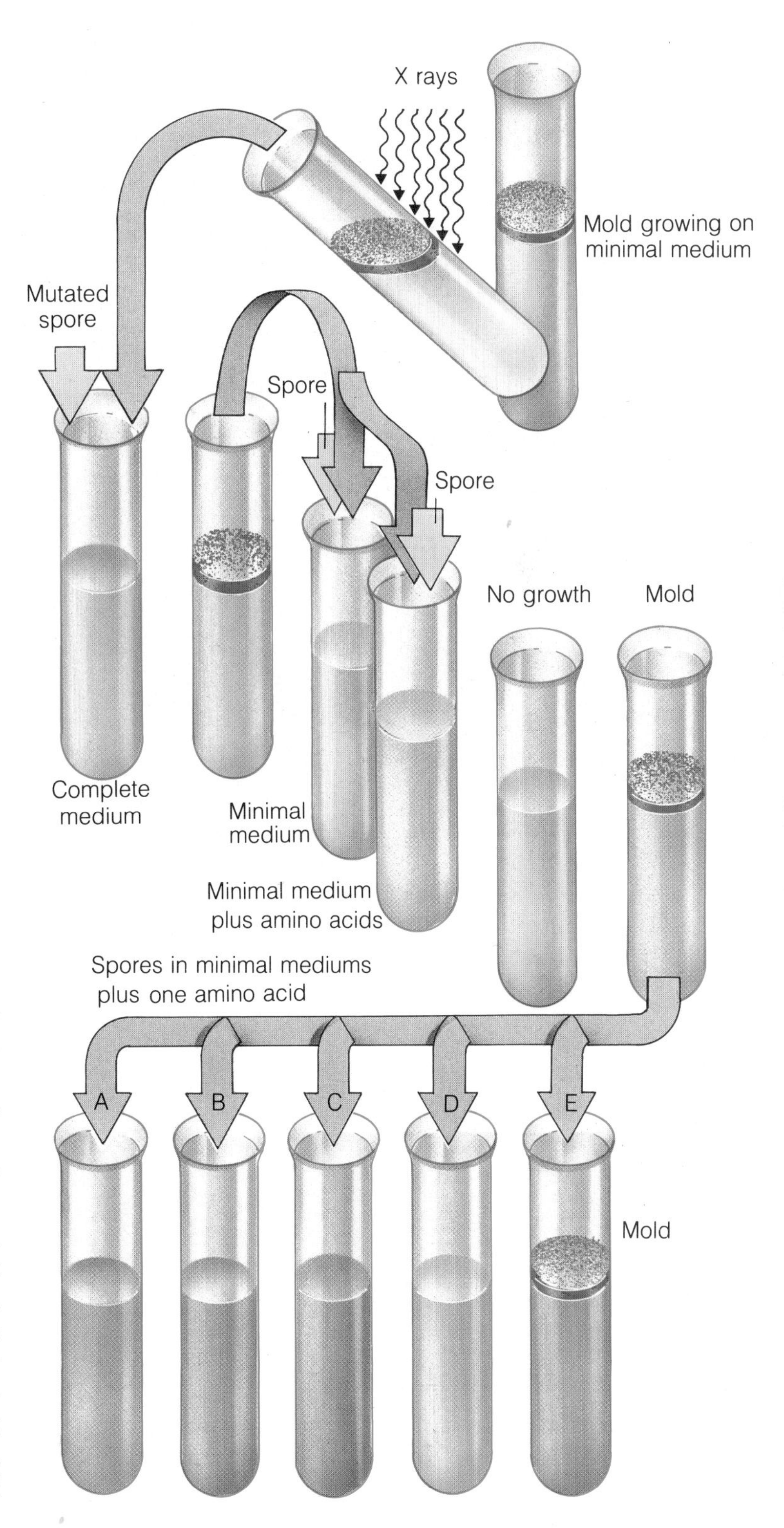

arginine itself. Beadle and Tatum realized that each step was controlled by a single enzyme and therefore that each gene controlled the production of one enzyme. The one gene–one enzyme hypothesis has been modified over the years, but it still dominates genetic thinking.

Now it was clear that many genes worked by controlling the cell's chemical reactions by means of enzymes. The gene must somehow contain the code in DNA that determines the structure of the enzyme whose production it controls. Many human genetic diseases are related to defective enzymes produced by mutant genes—at least 150 different disorders caused by enzymopathies have been recorded. This knowledge has allowed scientists to devise different strategies to help people afflicted with such disorders. Sometimes the missing enzyme itself can be replaced by transfusions; sometimes the missing enzyme step can be by-passed altogether by supplying the product of the reaction. Ultimately it may be possible to replace the defective gene and restore the cell's economy completely.

While these genetic experiments were coming to a climax, biochemists had been busy working out methods of determining the exact order of amino acids in a protein. This sequence is called the primary structure of the protein and it is directly related to what a gene does. The disease sickle-cell anemia was known to be caused by a change in the oxygen-carrying molecule hemoglobin in red blood cells. A single amino acid was implicated, but so basic a change also showed that the primary structure (the exact sequence) was determined by the gene. The idea was established that genes were lengths of information determining the exact order in which amino acids connect to form protein.

One of the most mysterious processes in biology is how a fertilized egg can develop into the hundreds of different kinds of cells found throughout the body. An unbelievably complex sequence of events must be carefully orchestrated to get every cell doing the right thing in the right place at the right time. Discovery of controller or regulator genes gives clues to how such differentiation might occur. When a single activating substance is produced, several different genes on different chromosomes may be switched on or off

Sickle cells (left), *shown here next to normal red blood cells* (right), *are a hereditary trait and result from a change in normal hemoglobin, leading to anemia. The disorder occurs mainly among black people, and affects about eight per cent of the black population of the United States. Patients with sickle-cell anemia have recurrent bouts of pain and fever, which generally subside in a few days.*

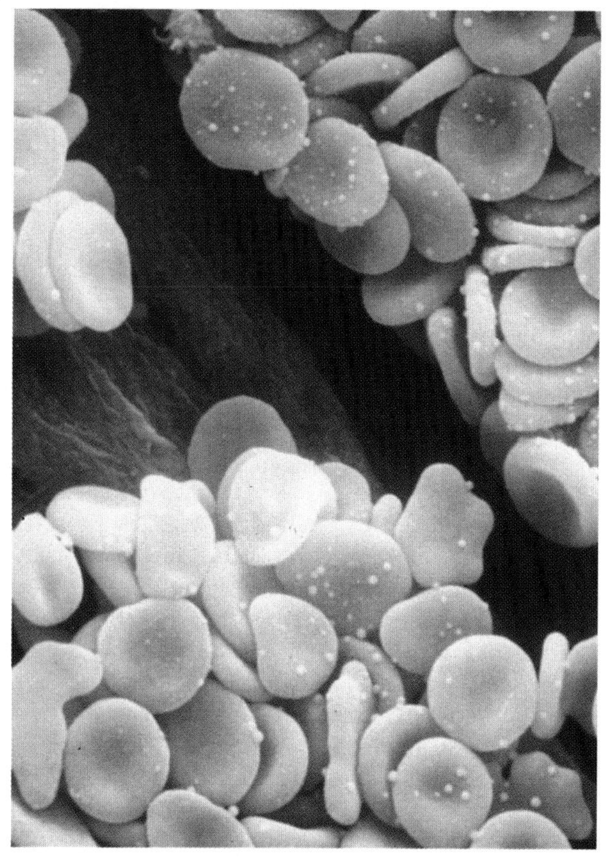

simultaneously. A large number of genes could be regulated by a single signal, causing sudden dramatic changes in the internal economy of the cell. Such changes are thought to be the basis of differentiation.

So far scientists had established that genes seemed to specify the exact order of amino acids in proteins but they had not discovered how the information contained in genes was converted into proteins. It was not obvious how a code consisting of four bases in nucleic acid could be read out into the twenty different amino acids that make up proteins. Some sort of code of different combinations of bases must signify which amino acid was required. What was the code?

The Message Gets Through

A useful clue to how information passed from gene to protein was the fact that the DNA was locked in the nucleus of a cell but all the protein synthesis occurred in the cytoplasm outside the nucleus. Proteins could not be made directly from DNA. Some sort of intermediate message must carry the code from the nucleus to the cytoplasm.

A candidate for such messenger activity was ribonucleic acid (RNA), which resembled DNA in structure and was abundant in the cytoplasm of cells making large amounts of protein. What is now called messenger RNA (mRNA) is synthesized in the nucleus as a direct copy of a gene and then exported. But there was still a major question: how did a string of bases make a string of amino acids? Chemical studies showed that there was no direct way the amino acids could be attached to mRNA. How was the code decoded?

Francis Crick, who had described the structure of DNA, also suggested that some sort of adaptor would be required. Such an adaptor would hold the amino acid at one end and attach to the message at the other end — just like adaptors for electric

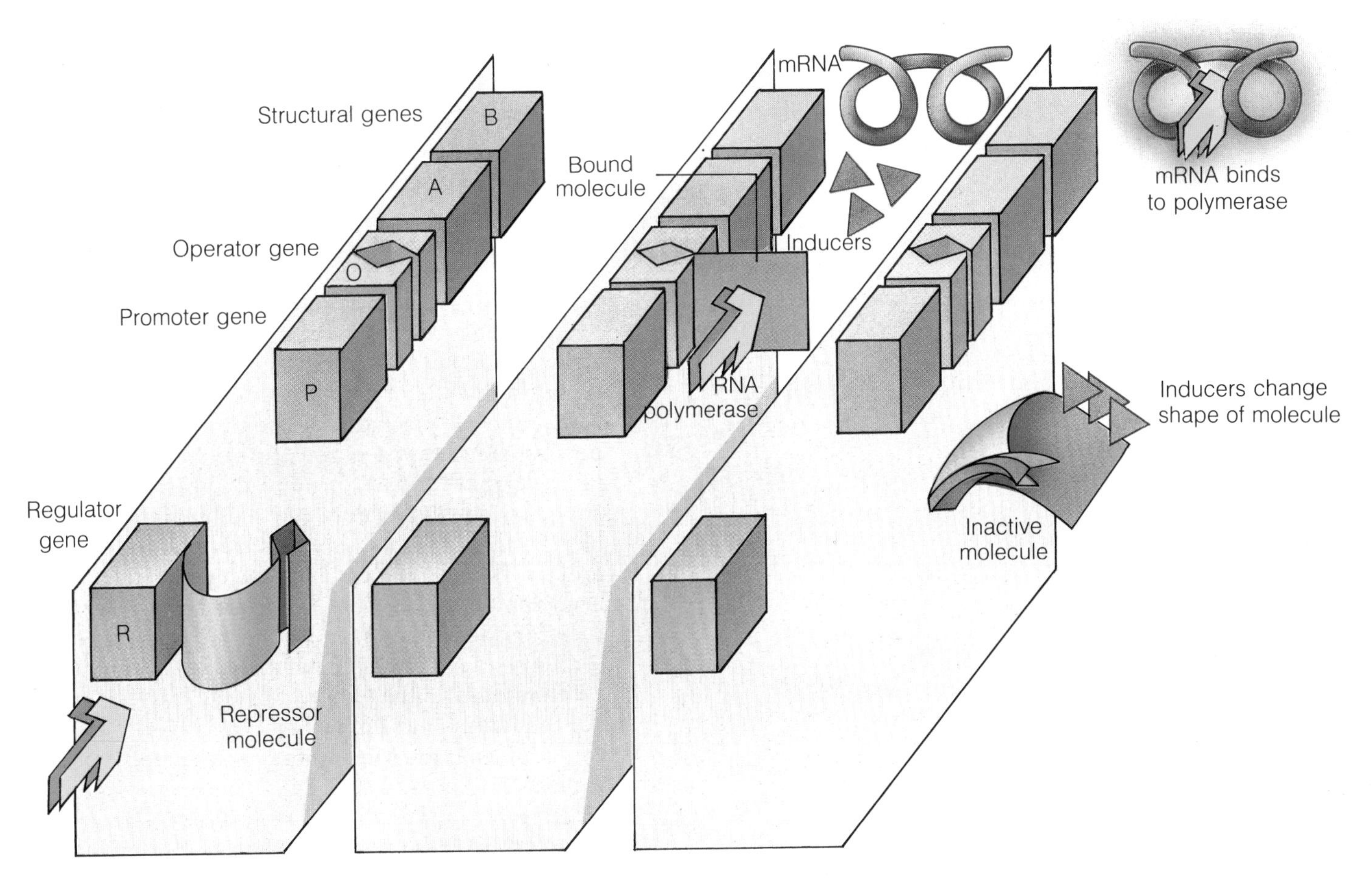

Protein production is controlled at several sites on a DNA molecule. A promoter (P) sends out RNA polymerase to promote the synthesis of messenger RNA (mRNA). A repressor (R) releases molecules to stop mRNA production. They do so by binding to the DNA at a site called the operator (O) and blocking the passage of RNA polymerase to the structural genes (A and B) which synthesize mRNA. Enzymes called inducers bind to the repressor molecule at the operator and change its shape so that it detaches itself. This allows the RNA polymerase to go to the structural genes. They synthesize mRNA, which attaches to the RNA polymerase and travels to the ribosomes, which make proteins.

razors which fit into the lighter socket on a car's dashboard and also take a standard electric plug.

After a period of intense speculation and experimentation the unexpected discovery was that there are three different kinds of RNA involved in making proteins. The bulk of the RNA transported from the nucleus turned out to be ribosomal RNA, which carried the information necessary for the synthetic machinery required to assemble the amino acids. Then amino acids were found attached to small pieces of another type of RNA before they were ever attached to the protein factory. These carrier RNAs were Crick's adaptor molecules and are now called transfer RNA (tRNA). Real messenger (mRNA) proved hard to find because it was very unstable and rapidly broken down.

Now that all the pieces of the protein-making machine were known it became possible to see how they worked together. Ribosomes were the assemb-

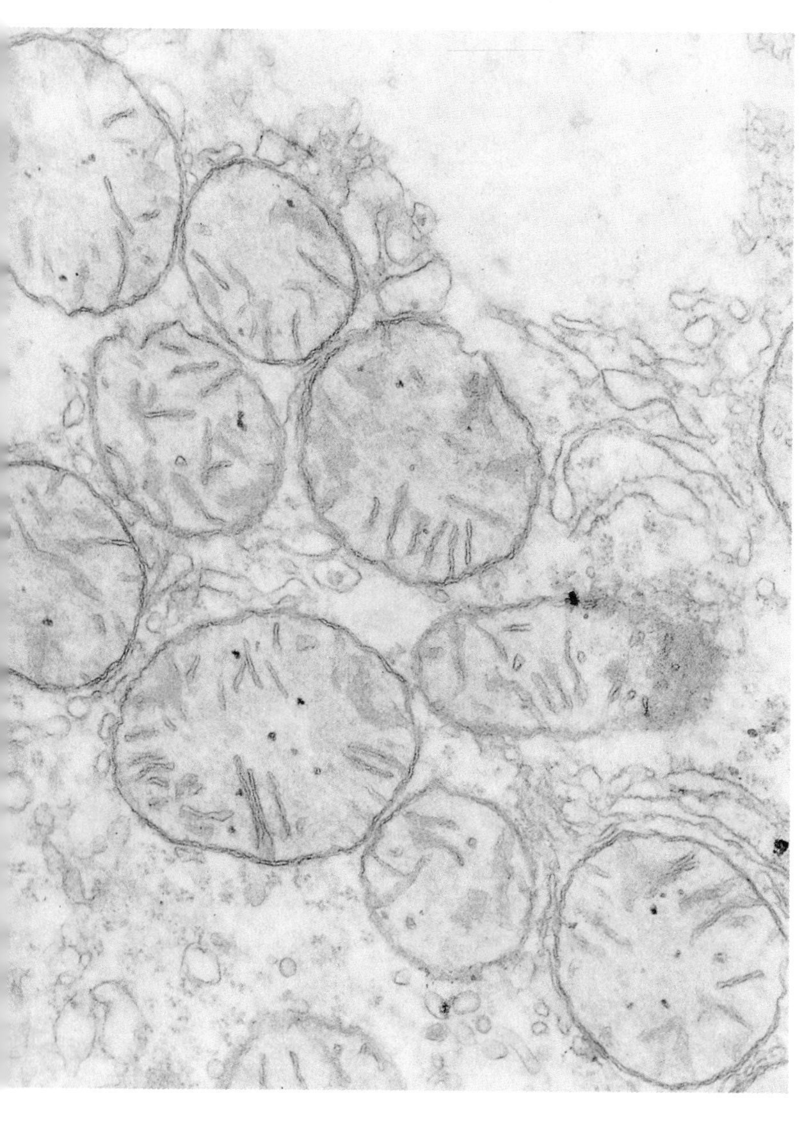

Mitochondria are the only structures in the cell, apart from the nucleus, which have their own DNA and synthesize their own proteins. Properties such as these have given rise to speculation that mitochondria were once independent organisms which were "eaten" by cells and developed so that they could live inside cells without harming them.

Albinism, a condition in which a person lacks pigment in the hair, skin and eyes, occurs when two genes come together from parents who may have normal coloring but carry the recessive gene for albinism.

ler framework and progressed along the mRNA strand. Transfer RNA, shaped like a three-leafed clover, fitted into a slot on the ribosome with the amino acid attached to its stem. Special enzymes coupled each new amino acid into the growing protein chain, and then the ribosome moved on a notch to the code for the next amino acid.

But still the elusive code had not been broken. Scientists then tried to produce an artificial mRNA to see what amino acid chain resulted. They used a string of uracil molecules, one of the four bases found in RNA, and added it to a complex mixture containing all the parts of the protein factory. A new protein composed of a string of the amino acid phenylalanine was produced. Messages of adenosine (poly A) produced chains of the amino acid lysine. By this simple trick the code was broken. Because the code for each amino acid had been shown to consist of three bases in a row, the code for phenylalanine must be UUU and the code for lysine, AAA. By 1966 the full code was known.

Interestingly the code turned out to be "degenerate," and more than one triplet of bases could indicate the same amino acid. Three special triplets UAA, UAG and UGA were stop signals which told the factory where to finish a protein. Although all these experiments were performed in bacteria, the universal nature of DNA and the similarity of its structure throughout all living organisms seemed to indicate that the genetic code was all but universal. One of very few exceptions was the code used by mitochondria, which had their own DNA and protein factory. For some reason their code differed by using two triplets for different amino acids.

The unraveling of all the parts of the protein factory revealed a complex system of interactions. Each transfer RNA had to have a code-recognizing end, which bound to the correct triplet on the message. Therefore there had to be special enzymes which made certain that the correct amino acid was attached to a tRNA bearing that amino acid code. Enzymes were needed to stitch the amino acids together and to make the ribosomes, tRNAs and message. Although the factory is complex, it seems to function remarkably accurately to produce exactly what the genetic code directs. The average protein contains about 400 amino acids, so its

message must be about 1,200 nucleotides long. How was this message contained and maintained in the molecules of DNA?

The Double Helix

Watson and Crick's model of DNA showed how it could consist of two strands of bases or nucleotides linked together as two mirror-image chains. The four bases were really two complementary pairs — adenine always partnered thymine, and guanine partnered cytosine. The bonds holding the strands together were just weak enough to allow the chain to be pulled apart, but also strong enough to let the two halves reassociate readily. This self-sticking property of half a DNA molecule was useful in identifying specific genes.

The complementary nature of DNA strands also explained how the genetic code could be correctly duplicated time and again. When the two strands were pulled apart a new copy exactly corresponding to the missing partner was made. When DNA, and therefore genes, were made each new copy contained one strand from the old molecule and one newly made strand.

Many of the mysteries of heredity could now be understood. Chromosomes are long threads composed of strands of DNA and genes are short segments in these threads. Most of the time DNA is faithfully copied so that two identical genes are formed, but if a small mistake is made then the wrong base will be inserted in the sequence. The protein factory has no way of detecting the change so a message containing the altered code will be produced and translated into an abnormal protein. Sometimes the resulting protein can perform normally, but often it cannot and some vital function inside the cell is disrupted.

Mutations (changes in an inherited characteristic) must result from such tiny changes in the code. Often these are random accidents, but physical

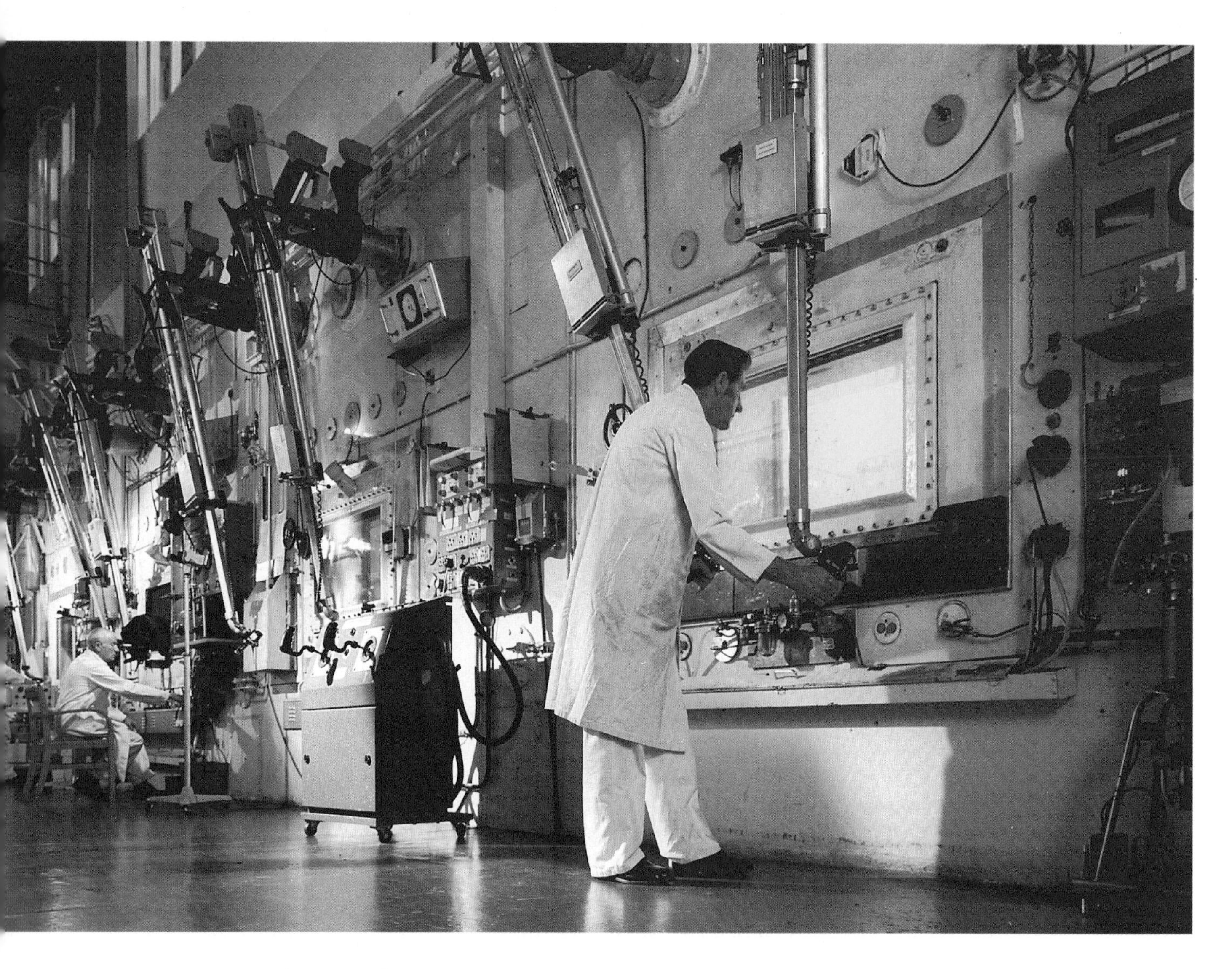

agents such as radiation and cancer-causing chemicals also produce them. Radiation, whether atomic or from sunlight, may damage single bases or knock out a whole short sequence (which is then known as a deletion) from one DNA chain. However, radiation is universal and it is a great advantage to have a mechanism for repairing damage to DNA, or hereditary information would be rapidly destroyed. There is, in fact, a set of enzymes that repair DNA. These clip out damaged bases or fill in gaps so that the original sequence is restored. The importance of these enzymes is suggested by the fact that cells have at least fifty different ones ready for use at all times. Of the three billion base pairs in a typical mammal's sex cells, only about fifteen change in a year.

Treatment of Cancer

One strategy for destroying unwanted cancer cells involves poisoning them by damaging their DNA so that they cannot reproduce. Both radiotherapy and chemotherapy damage genetic material. Many chemotherapeutic agents can be incorporated into the DNA strand and cause false messages. Others link the two complementary strands together so that they cannot separate and be copied. But such treatments may also damage normal DNA and lead to sterility or paradoxically to the formation of new cancers if the patient survives.

The discovery of the genetic code and the protein factory represents the single most important advance in biology ever made. Despite this there are still many mysteries surrounding the information contained in the code. The mechanism that switches parts of chromosomes on or off is poorly understood, for example.

The form of the chromosomes within the cell between mitotic events is also a puzzle. One suggestion is that they may be attached to the nuclear membrane by a kind of hook. Parts of the

The most stringent measures for protection against radiation, such as eighteen-inch plate glass, are essential to prevent contamination. Radiation can damage DNA and lead to mutated cells.

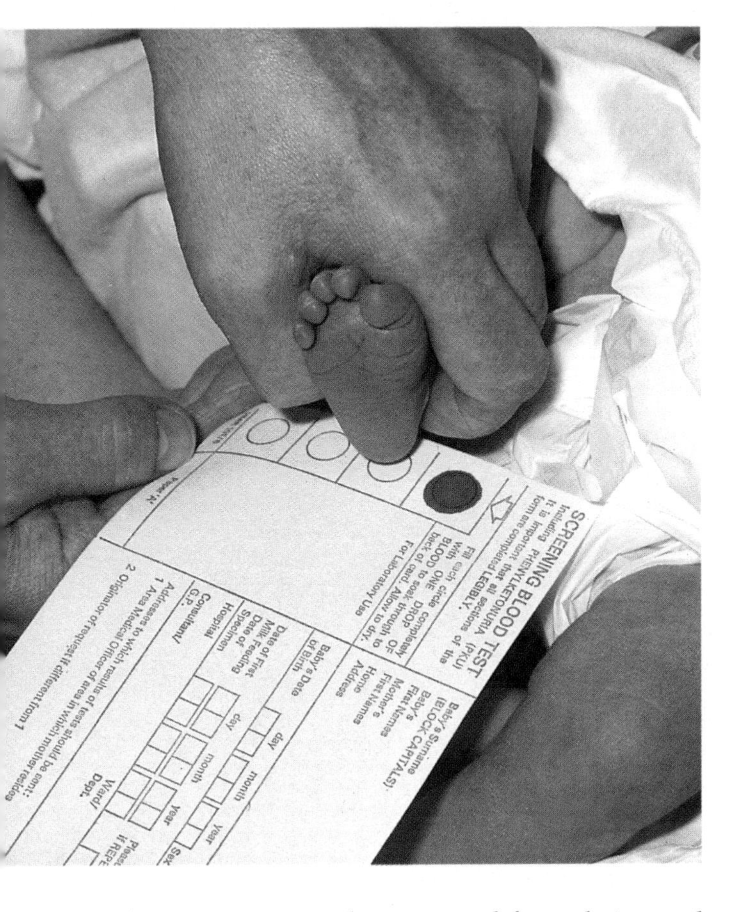

chromosomes may be unwound from their spool and the DNA stretched out into a series of loops. Control proteins probably bind to specific control genes making the DNA unfold further so that the enzymes that produce mRNA can attach to the strand. These enzymes, called RNA polymerases, make many mRNA copies of the opened-up genes. These are then sent to the cytoplasm to produce proteins. Off-switch proteins, called repressors, may block the action of the RNA polymerase or cause the DNA to coil itself into a shape that is unreadable.

Perhaps the most pressing issue in biomedical science today is how these control genes and the proteins which interact with them control cell growth and division. Many scientists suspect that it is these mechanisms which constitute the biological basis of cancer. Understanding them may make possible the control of deranged malignant cells and offer a cure for the disease.

A small sample of blood is taken from a baby to test for cretinism (above left), *which may result from a lack of the enzyme essential to the normal breakdown of the amino acid phenylalanine. An abnormal accumulation of this amino acid in the brain leads to mental deficiencies. The condition results from the homozygous presence of a recessive gene (pp). Physical agents can also cause changes in the genetic code and lead to mutation; for example, smoke in the lungs can be cancer-causing, forming carcinomas in the lung tissue* (above).

Chapter 4

Inborn or Acquired?

Many aspects of life, including life style, education and diet, are just as important to the individual person that you are as your genetic constitution. After all, genes cannot exert their effects without an environment in which to operate, and environmental factors act differently on different people. Conversely, the environment may be the same, but people are still individuals — their different genetic constitutions make sure of that. The interaction between heredity and environment is thus both universal and complex, and the subject can be examined at several levels: from the broad concepts of the evolution of life down to individual problems (or advantages) that one child may inherit.

Environment — Mold or Blueprint?

At one time it was thought that environmental factors were the determining influences on heredity. The nineteenth-century French evolutionist, Jean Baptiste de Lamarck, taught that acquired traits were predominant. He thought, to repeat the example, that giraffes had acquired long necks because their ancestors stretched upward to feed on high branches and lengthened their necks, a characteristic which in turn they passed on to their offspring. If this were true then skill in painting, or dancing or playing the piano could be handed down to another generation — but it is not.

It is possible to consider the environmental effects on heredity from a different viewpoint. If having a long neck allows an animal to get more food, that animal will probably have more offspring, which will have long necks because their parents had. There is an advantageous interaction between an inherited feature (neck length) and the environment (the high leaves on the trees). If there were no tall trees for food, there would be no advantage to having a long neck, and the animal with that particular abnormality would be no more successful than another. In consequence the trait might simply breed itself out again.

The woodpecker finch uses a twig to prod at a hole in a treetrunk and pull a grub to the surface. The bird is one of a group discovered by Charles Darwin in the Galapagos Islands, in the Pacific Ocean off the coast of South America. In this competition-free environment, the finch is adapted to occupy a particular ecological niche. It was by observing such creatures that Darwin got his ideas about evolution and natural selection.

Six antibiotics were applied to a bacterial culture; the result shows that the bacterium is resistant to three of them. Such resistance develops through subtle changes in genetic makeup of microbial cells.

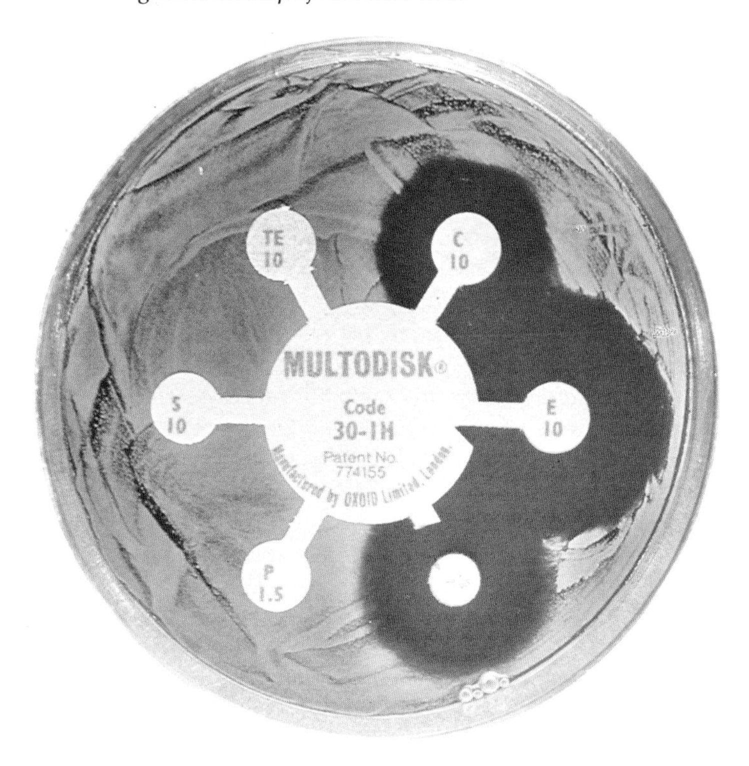

However, there are two dramatic examples of the effect the environment can have on genetics: drug-resistant bacteria and pesticide-resistant insects.

Before antibiotics were in wide use, most bacteria were highly susceptible to their effects; the age of the miracle drug seemed to have dawned. All the same, an occasional bacterium sometimes appeared through spontaneous mutation that was resistant to the normally devastating effect of one or other antibiotic drug. In former days there was no particular advantage in having the gene for antibiotic resistance, and the bacterium would have bred the trait out again in future generations. But in the face of the antibiotic revolution this resistant gene gave a bacterium two considerable advantages. Not only could it survive, but it found itself in an environment free of other organisms competing for food, and nutrition was limitless.

Not surprisingly, such organisms increased rapidly in number and became very widespread. Yet many resistant organisms are successful only where antibiotics are in constant use, such as in a hospital. Overall, the original nonresistant forms of bacteria remain more prevalent, indicating that they are slightly better adapted to life without antibiotics than their drug-resistant cousins are to life with them. This emphasizes an important point regarding the interaction between heredity and environment: a small, transient change may yet have profound effects that work either for or against a particular gene.

Many bacteria have gradually built up resistance to a whole range of antibiotics, and in order to combat this growing trend new antibiotics are kept out of general circulation and reserved solely for use against organisms resistant to all other drugs.

This resistant trend is occurring not only among organisms in hospitals but has also encroached through indirect effects on the human food chain. Antibiotics are often used to supplement animal feeds because they can enhance the animals' growth rate. But organisms acquire resistance to them and can then spread to humans in contaminated food. Some experts maintain that it would be a wise policy to ban the use of antibiotics in animal feeds to further limit the spread of bacterial resistance.

At one time or another, you may perhaps have wondered why a physician makes a point of instructing patients to be sure to finish a course of antibiotics once started, even though the infection for which the drug has been prescribed has cleared up before the course is completed. In view of the adverse publicity that from time to time surrounds antibiotics you might imagine it would be a good idea to stop taking the tablets as soon as possible. Unfortunately the opposite is true. Even when the drug has helped the body to overwhelm the infection, there may be a few bacteria in the system which are still alive although they may not be adversely affecting health, Finishing the course will probably kill these slightly stronger strains as well.

The "Super" Insects

Pesticide-resistant insects represent a major threat to both world health and economy. The *Anopheles* mosquito, which carries malaria, is a good example. The insect—and thus the disease it carries—was formerly controlled by DDT which used to kill it. Some mosquitoes, however, had just enough resistance to survive and breed. A few of their offspring were resistant to even greater levels and

A fanciful tale by Rudyard Kipling tells How the Camel got his Hump. *The camel was an idle creature; it spent all day eating and refused to work. As a punishment, a genie produced a hump for the camel's back in which it could store enough food to enable it to work for three days without eating. Ever since then, so the story goes, the camel has worn a hump—falsely suggesting that acquired traits are inherited.*

Following a battle in World War II, men of the American 35th Infantry Division remove casualties for hospitalization. Breakthroughs in the development of antibiotics at that time enabled better treatment of wounds and injuries. The later widespread use of antibiotics led, however, to a gradual build-up of resistant strains of bacteria, resulting today in limitations to the use of new antibiotics.

survived the next spraying, and so on. Because at first there were only a few survivors, it appeared that the mosquito had been conquered. But gradually the mosquitoes became resistant to maximum doses, and their numbers began to expand rapidly despite regular spraying. With the increase in numbers came the reemergence of malaria in areas where it had been eradicated. And, inevitably, the malaria parasite itself has also developed resistance to many of the agents used to treat it, much as bacteria have acquired antibiotic resistance.

So how did the first few mosquitoes resist the initial spraying? Perhaps they were hidden and exposed to only low doses of pesticide, and a gene or genes in their makeup gave them just enough resistance. This gene would then be passed on to offspring backed up, perhaps, by strong genes from another parent. It is intermittent low doses of a substance that encourages resistance. Eventually a few totally resistant insects would be born which increased rapidly — with no competitors.

A similar pattern is emerging with regard to another insect, the head louse. This bug, not at all common outside Europe, lives close to the human scalp and lays its eggs on the hair; the eggs are known as nits. Head lice seem to have a preference for clean hair and are fairly common in schools, where the lice can easily be transmitted from head to head. Parents, in their eagerness to remain anonymous, treat their children with special head-lice shampoos which are weak and, unlike head-lice lotions which contain strong pesticides, can actually encourage resistant strains of louse to evolve in just the same way as the supermosquito.

Small variations in local environment, such as temperature, have a similar effect. The American evolutionary geneticist Theodosius Dobzhansky has spent many years studying the effect of such minor variations in ecological niches on the

The only survivors of an initial spraying of insecticide (below left) *are those insects which, as a result of a genetic change called a mutation, possess a gene that makes them capable of withstanding the insecticide's effects. These insects then pass on this "strong" gene to their offspring. On each consecutive spraying of the same insecticide, a greater number of insects have developed resistance. The insecticide eventually ceases to be effective in killing the new resistant strain of insect—in this example, the malaria-carrying* Anopheles *mosquito* (bottom), *seen here sucking blood from a human host.*

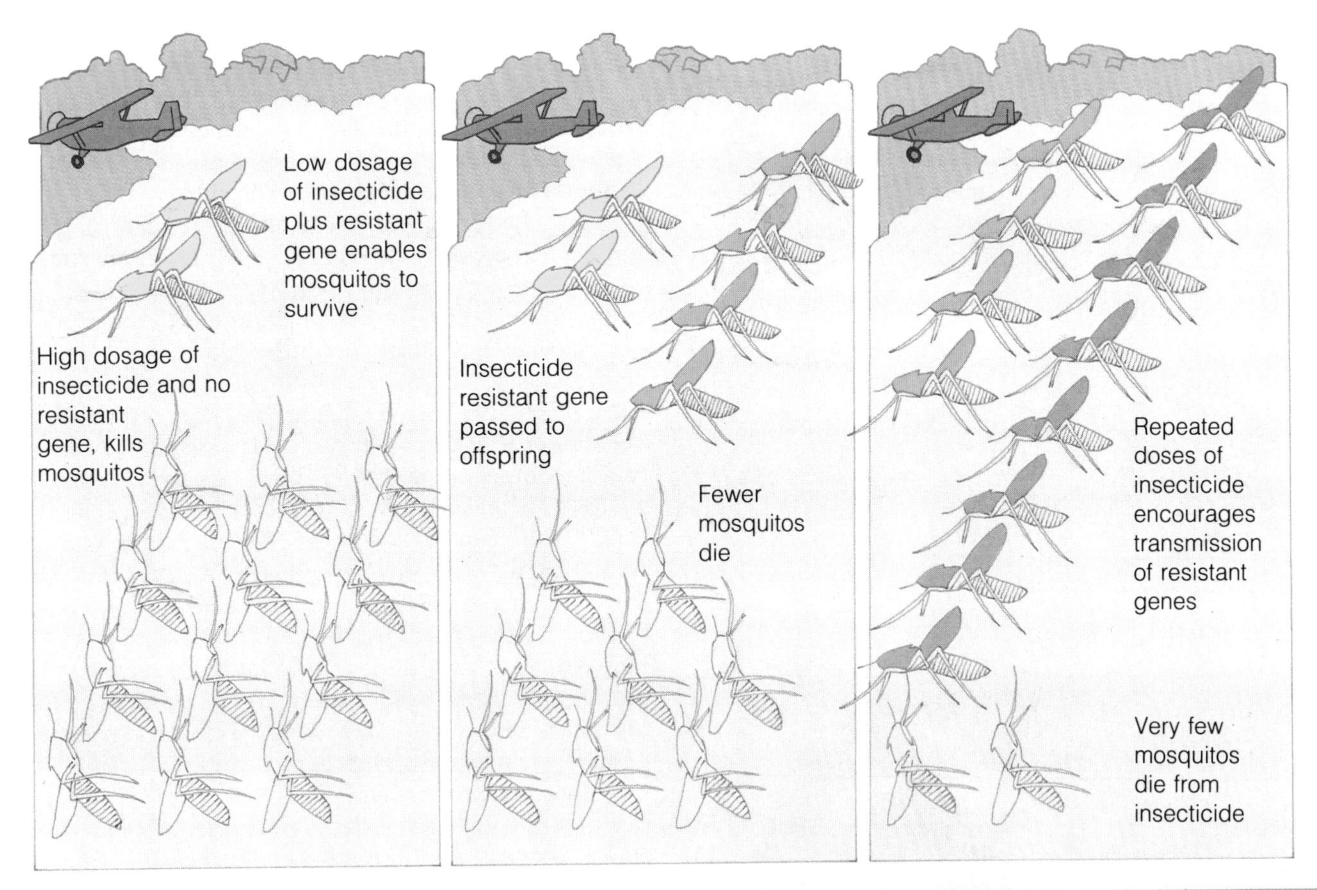

development of new species of fruit flies in the wild. Many closely related but distinct species have developed, probably to exploit small but specific differences in the local environment.

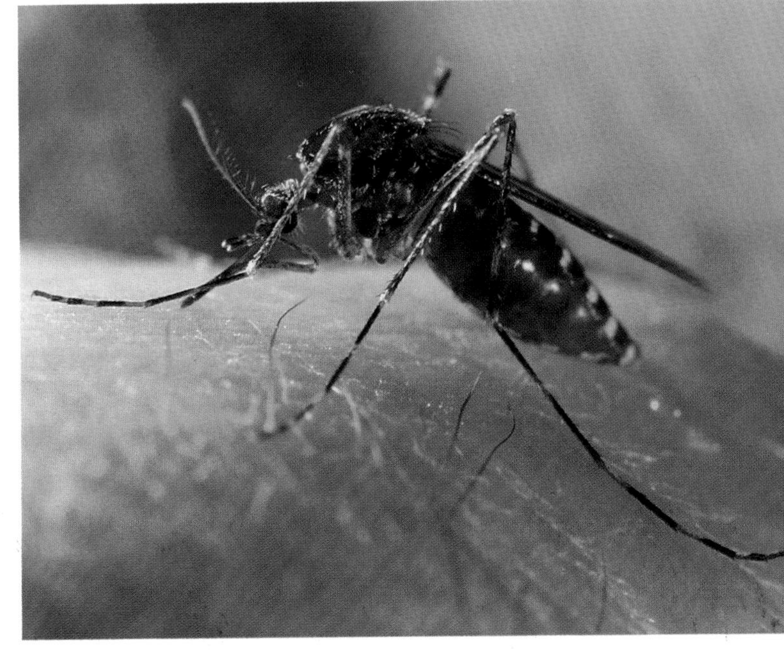

The Theory of Natural Selection

Some individuals survive better in one set of surroundings than the other members of their species, and the long-term effect of this is known as "natural selection." Darwin's theory of evolution by natural selection was formulated before the laws of genetics were discovered. He was unable to describe how change took place. The doctrine of "survival of the fittest" was often taken to suggest intense physical conflict between members of different groups or species. It is now known that evolution is the process by which individuals, with genes that give them an advantage in a particular environment, produce offspring that are more successful than their competitors for food and

Animals such as this reptile (below) *were studied by Darwin during his voyage on H.M.S.* Beagle. *Darwin's ideas laid the foundations of the theory that individuals with genes that enable them to survive in a particular environment have offspring which are better adapted to that environment. The dinosaurs* (bottom) *eventually became extinct, possibly because they could not adapt to environmental changes.*

living space. Their genes gradually increase, and at a certain stage the species becomes so well adapted to its ecological niche that any further change would be a disadvantage. Long periods of stability ensue, because the fittest are now the most common as well as the most successful.

Rapid changes in species probably occur during times of great climatic change, such as in an Ice Age. Following the onset of Arctic cold, among those animals already better equipped to survive because they have thick fur and a resistance to cold, some may appear with a mutated gene which gives them strong teeth. This means they can chew the vegetation that grows in the cold climate. A new trait becomes established and, as the most successful "model," continues to breed a species totally suited to that cold desolation. Adaptation would be of no advantage thereafter until or unless the environment again altered dramatically.

Evolution is therefore the result of interactions between changing genes and changing environments. Mutations (which are usually harmful) are necessary for adaptation to occur. If there was no genetic variation, a group might simply die out if the environment changed.

Many scientists think that this is what happened to the dinosaurs. Although very successful — so much so that they dominated the world for 150 million years — a sudden dramatic fall in the Earth's temperature destroyed their environment faster than their genes changed. Suddenly it was an advantage to be a small, furry animal like a primitive mammal. One theory is that this environmental change was caused by a collision between the Earth and a large meteorite or asteroid which spread so much dust into the atmosphere that sunlight was partly filtered out and the temperature dropped, remaining low for years afterward. Scientists now tell us that much the same conditions would probably follow a nuclear war, and that the "nuclear winter" is likely to rapidly wipe out most survivors of the initial holocaust. Perhaps the Eskimo (Inuit) people would be the only race genetically and traditionally able to face this challenge in any way.

Even in the everyday ordinary world, however, there are deleterious genes that convey an advantage to those who carry them. An example of this is the gene that causes sickle-cell anemia, which occurs among African people; Mediterranean peoples suffer from a similar gene defect which causes thalassemia. Both these genes are recessive, and if a person inherits two copies (that is, if he or she is homozygous) there is a high risk that fatal anemia will develop because these genes affect the formation of hemoglobin, the oxygen-carrying protein inside red blood cells. Under normal circumstances this potentially dangerous gene should become less and less common because some of its carriers would die before having children. However, both these genes are common in parts of Africa, and in Italy about one in every hundred children is born with thalassemia. Why should such a harmful gene be so common?

The reason is that, in specific circumstances, people who have inherited one copy of the sickle-cell or thalassemia gene (that is, they are heterozygotes) have an advantage over people with no copies of the gene. The slight abnormality of their red blood cells actually protects them against malaria, a disorder caused by parasitic invasion that destroys normal red cells. Heterozygotes tend to have more children than people with fully normal red cells — but only in an environment where malaria is common. When the disease is eradicated, as in Italy, or a person moves to a malaria-free zone, such as the United States, then the advantage ceases. Gradually the abnormal gene declines in frequency as the reproductive advantage of having it disappears.

In some parts of Italy such as the Po Valley, the thalassemia gene strongly influenced social behavior between the fifteenth and eighteenth centuries. A woman with one thalassemia gene was often moderately anemic and had a characteristically pale face and dark-eyed look: such women were widely sought after. The look was also considered the height of femininity because it was common knowledge that such women tended to have larger families. Of course, nobody realized why this was so.

Similar examples of environmental effects on heredity were seen when Europeans began to colonize North and South America. The local population had never been exposed to measles or smallpox, both common infections in Europe. As a

Dark eyes and a pale complexion are characteristics of thalassemia, an often fatal hereditary anemia mainly found in people of Mediterranean origin. People who are carriers have greater resistance to malaria.

Sickle-cell anemia first became established in environments favorable to malaria-carrying mosquitos. People with sickle-cell disease (A) have collapsed, sickle-shaped red blood corpuscles that clog the capillaries, often proving fatal. In people who are carriers of the sickle-cell trait (D), less than one per cent of the red corpuscles are abnormal. These individuals do not die from malaria, like those (B) with no sickle-cell trait, nor do they suffer from anemia, but they do transmit the sickle-cell to their offspring. With a move to an area with a low risk of malaria, the sickle-cell gene is eventually removed. (C).

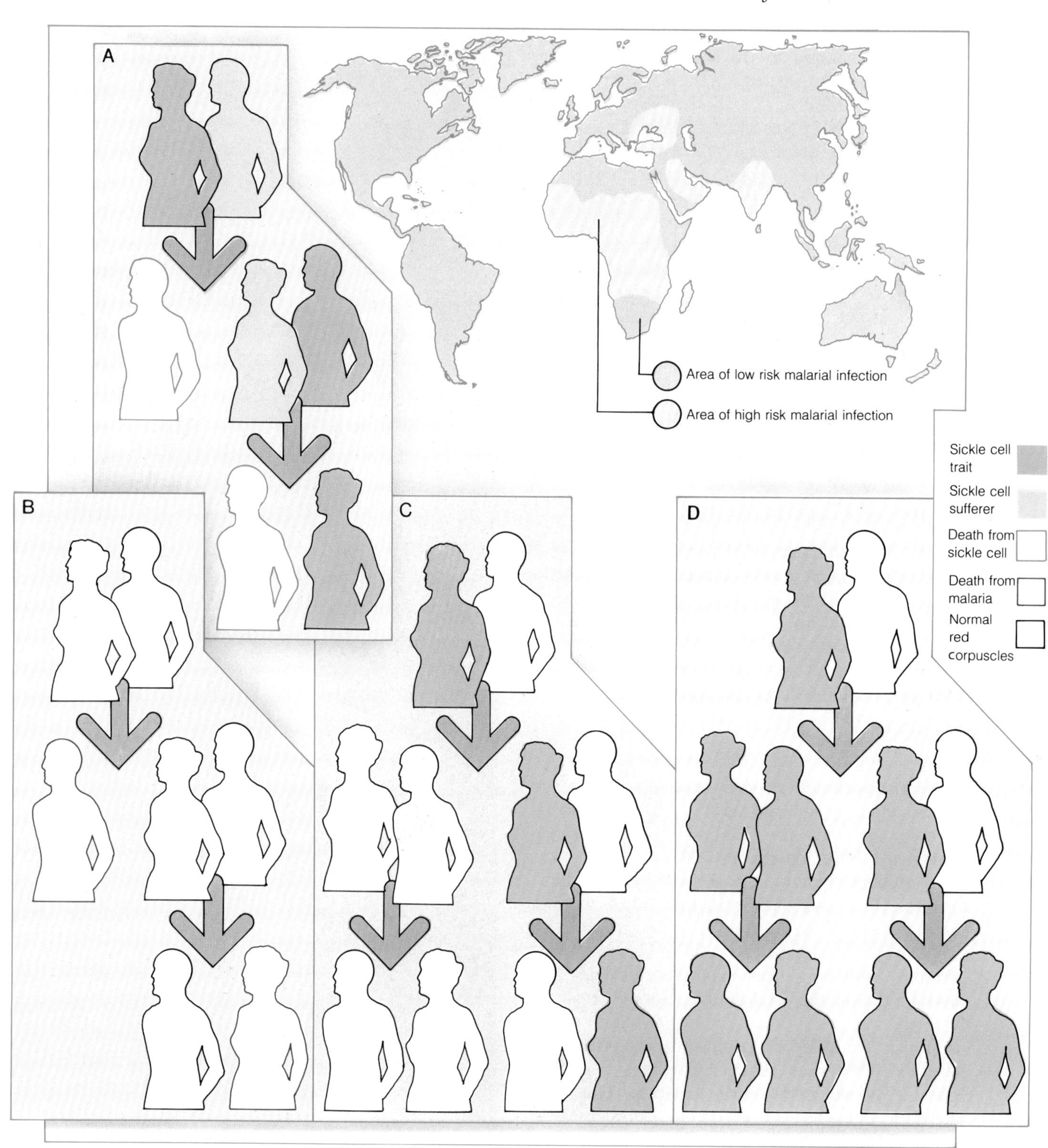

People who live in mountainous regions are usually adapted to the thin air of high altitudes by an increase in their lung tissue. The characteristic is not, however, passed on to their offspring.

result there was little resistance to these viruses and many people — especially children — died. Resistance to a particular infection can build up in a population over many years in much the same way that mosquitoes became resistant to DDT. After a few generations those with resistance will have survived and have had children, thus spreading their genes through the population.

Interestingly, there is a suggestion that the indigenous Americans may have had partial revenge on their conquerors. Syphilis appeared in Europe at about the same time as Colombus returned from America and for two or three centuries ran rampant. The disease seems to be less severe now than in the past, possibly because the population has built up a genetic resistance.

A Modern Environment

Attempts can be made to treat or prevent the effects of inherited diseases. These constitute a special kind of interaction between heredity and environment in that they are a deliberate attempt to alter the environment and so lessen the effect of unfavorable traits.

A good example of this type of intervention is the treatment of an uncommon condition called phenylketonuria (PKU). About one child in 10,000 is born with PKU, which results from having a specific recessive gene. The PKU gene causes a failure in the production of an enzyme that normally converts the amino acid phenylalanine into the amino acid tyrosine. Because this enzyme is not present, an excess of phenylalanine builds up in the body and the child's developing brain is damaged. The result is that PKU children become severely mentally retarded.

If the disorder is detected very early after birth, a diet free of phenylalanine can be instituted and the damaging build-up of this amino acid prevented. PKU children who have had this special diet grow up to have IQ scores very similar to their unaffected brothers and sisters. By the time the child is in his or her teenage years and the brain has fully developed, it is possible to relax the strict diet

Many indigenous Americans died when they contracted diseases introduced by European colonists. Never having been exposed to such diseases, local people had little or no resistance to them.

without harmful effects. All children are now screened for PKU by a simple test on a spot of blood taken at birth. This is a near perfect example of how the environment (in this case modern medicine and a special diet) can alter the expression of a hereditary trait.

Many similar interventions are now carried out. Children born with a genetic defect in their immune system formerly used to die of an infection in infancy. To prevent this, children may now be cut off from an everyday environment potentially lethal to them and placed inside a bubble chamber where they are exposed only to sterile air and food. This allows time for a search to be instituted for a suitable donor, for in some instances the deficit in immunity may be corrected by changing the child's internal environment by replacing the thymus gland or some bone marrow.

Recently a new group of inherited traits has been discovered. These traits are central to certain critical interactions between a person and the modern world, and involve genes that control the speed with which the body breaks down certain chemicals. Most of the chemicals are drugs or environmental pollutants, and the effects on a faulty metabolism can be profound. People who are slow metabolizers of particular medical drugs, for example, can develop unusually high levels of the drug in the bloodstream. Such high levels appear to dramatically increase the patient's chances of developing side effects. The speed at which a drug is eliminated from the body is also important in determining the length of time an effect persists. Although it is not yet possible to measure these gene effects outside the research laboratory, the time is coming when individuals will have their ability to metabolize drugs tested just as they have their blood grouped. Instead of a blanket dosage for everyone, the drugs can then be tailored exactly to each person's individual needs.

The discovery of this new group of inherited traits has opened other channels of investigation. There is a suggestion that a person's exact composition of drug-metabolizing genes determines how he or she breaks down the chemicals in cigarette smoke. Possibly some people produce high concentrations of especially dangerous metabolites and are therefore at much greater risk

Wilkins, Watson and Crick

The Molecular Model of DNA

The determination of the molecular structure of DNA—the chemical substance that is responsible for the passing on of hereditary characteristics—was a work of consummate brilliance that resulted fairly equally from the contributions of three scientists: Francis Harry Compton Crick, James Dewey Watson and Maurice Hugh Frederick Wilkins. Together they made the discovery that has been called the most significant advance in biology of the twentieth century, and were awarded the Nobel Prize for it in 1962.

Crick was born in June 1916 in Northampton, England, and graduated from University College, London. During World War II he worked on scientific instruments and weapons for the Royal Navy. But afterward he transferred to biological studies, arriving at the Cavendish Laboratory Medical Research unit in Cambridge in 1949.

Two years later, another researcher, Watson, also began to work there. Born in April 1928 in Chicago, USA, Watson studied viruses first at the University of Indiana and then at the University of Copenhagen, Denmark, before traveling to Cambridge.

There, Watson became interested in research into the structure of molecules using the X-ray diffraction techniques meanwhile being pioneered in London by Crick's almost exact contemporary, Wilkins. It was Wilkins—a New Zealander from Pongaroa who graduated in physics from Cambridge in 1938—who discovered the double helix shape of the DNA molecule and passed on the information to Watson and Crick.

Using that information, Crick and Watson worked to produce a complete model of the molecule that would correspond to all known factors and attributes. It became possible in 1953, only after Watson had discovered that the organic bases of DNA are linked as distinct pairs—adenine with thymine, guanine with cytosine. The finished model vividly showed how the DNA molecule could duplicate itself: each helical strand could individually serve as the pattern for the complementary formation of the other.

Crick remained at Cambridge until 1977, when he took a chair at the Salk Institute in San Diego, California. In 1985 Watson was Head of the Laboratory of Quantitative Biology at Cold Spring Harbor, Long Island, New York, and Wilkins was Professor of Biophysics at King's College, London.

Evidence suggests that certain people have a genetic predisposition to develop lung cancer (below). *As a result, smoking* (right) *can activate a particular enzyme which reacts with smoke and causes cancer.*

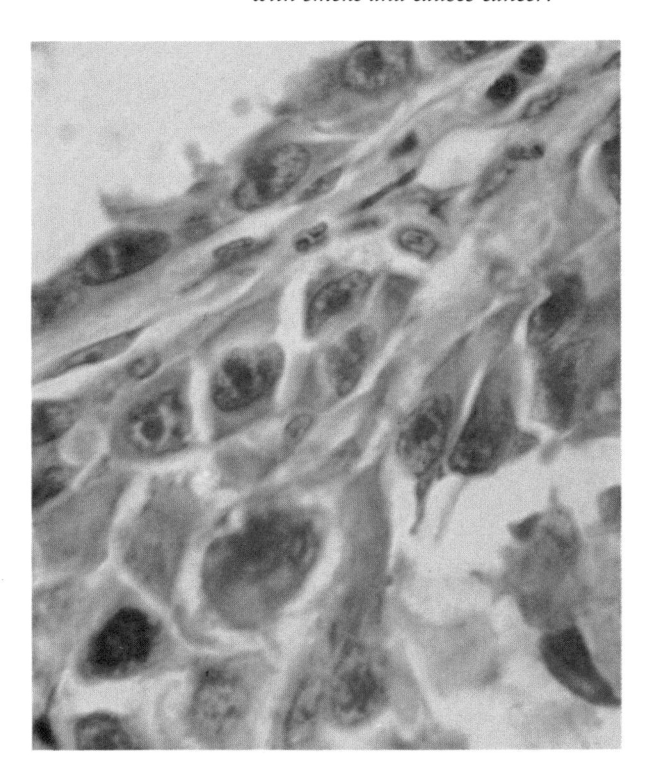

of developing lung cancer if they smoke than most of the population.

One epidemiologist (a scientist who studies the frequency and causes of disease) has suggested that if everyone smoked it would become clear that lung cancer was a genetic disease. As it is, it seems very likely that this cancer, and perhaps others, are caused by interaction of a susceptible genetic background and an environmental stimulus.

Studies of Twins

It is comparatively easy to detect environmental effects in certain disorders, but the problems really increase when attempting to distinguish between heredity and environmental components of many normal human characteristics. Why are you as tall as you are? Is it thanks to your mother or your father, or to the high protein diet you had while you were growing up? Why do you weigh as much as you do? Do fat parents tend to have fat children because of their genes, or because of the perhaps appalling eating habits with which they indoctrinate their children? Do you get your intelligence from your parents, or from the education you received from school and at home? Do highly motivated parents pass on genes for motivation or instil an urge to achieve through example? The argument is called the "nature versus nurture" conflict — "nature" refers to a person's genes, and "nurture" represents the environmental factors that surround them, whether these are social, educational or nutritional.

The interaction of these features is so complex that each is difficult to study separately, but much information can be gathered by looking at different kinds of twins. Three particular types of relationship are usually compared — identical twins raised together; identical twins raised apart from each other; and nonidentical twins raised together. Because identical twins have exactly the same genes, the effect of environment can be assessed by looking for differences between twins raised together and twins raised in separate homes. If both twins are the same when raised apart, then the inference is that environment is less significant than heredity. But if they are very different when separated, then individual traits must largely be ascribed to environment.

The outcome of such studies is important. Results can be used as a basis for important and expensive policy decisions. A government can manipulate the environment to a certain extent if traits turn out to be environmentally determined. Hereditary characteristics cannot be altered.

Unfortunately, studies such as these are open to abuse because they are interpreted by people who may impose their own personal prejudices on the results of the studies. One especially damaging argument that emerged a few decades ago claimed to prove that basic intelligence is determined purely, or in very large part, by genetics. It is now known that many of the studies which purported to prove this were either fraudulent or at least biased. Investigators involved chose to show the results which supported their own hypotheses concerning certain races, rather than those which pointed in a different direction. Many such studies were based, for example, on a comparison of the IQ test score results of groups of individuals making no allowance whatever for differences in education or social background. Apart from the strong probability that the results of IQ test scores reflect social

Tall individuals stand out noticeably above the crowd of characteristically short-statured Japanese. The younger generation, however, is now growing taller because of an increase of meat and dairy products in their diets.

background and not intelligence, the results could not compare the effects of environment on similar or identical genetic backgrounds. It is in this respect that twin studies become valuable.

The study of twins is limited by the fairly small number of twin children born each year, especially of identical twins who have been separated at or near birth. The number of identical twins reared apart becomes even smaller when a particularly uncommon trait or disorder is examined. Twin studies are most suitable for investigating common human features such as height, weight and IQ test scores. Common diseases can often be assessed, but for rare problems there simply are not enough affected pairs of twins available for investigators to reach any sound conclusions.

Twin Study in Action

A twin study is performed in one of two ways. Either the presence or absence of a particular feature is noted or a quantitive feature such as height is measured. For traits such as a disposition toward contracting a particular disease, pairs are scored as concordant if both twins have the disease or discordant if one twin is affected and the other is not. The result is usually given as a percentage concordant, that is to say, the number of twins that show the same result out of the total number studied. This sort of study has been carried out for many human diseases and not surprisingly some are more hereditary than others.

Rheumatoid arthritis, a crippling joint disease, occurs in both of identical twins only about ten per cent of the time. This is more often than the disease occurs in the rest of the population, but cannot be entirely due to genetic factors because if this was the case, every identical twin would have the disease if the cotwin did.

On the other hand, if one twin suffers from manic depression, then the cotwin has a sixty to seventy per cent chance of getting it too, whether the two are raised apart or together. This shows a strong genetic element. Note also that there is a subtle difference between genetic and hereditary traits. If a trait is caused by a particular combination of genes then it will probably occur in identical twins because they share the same genes. This means it is genetic. But if the combination of genes required is

Much of the evidence relating to the inheritance of intelligence comes from studies which correlate the IQs of persons of various genetic relationships. If two people have exactly the same IQ, the correlation coefficient is 1; smaller coefficients indicate less correlation. In this diagram, the vertical bands show the range of correlation coefficients, and the red pins mark the average value for each kind of pair studied.

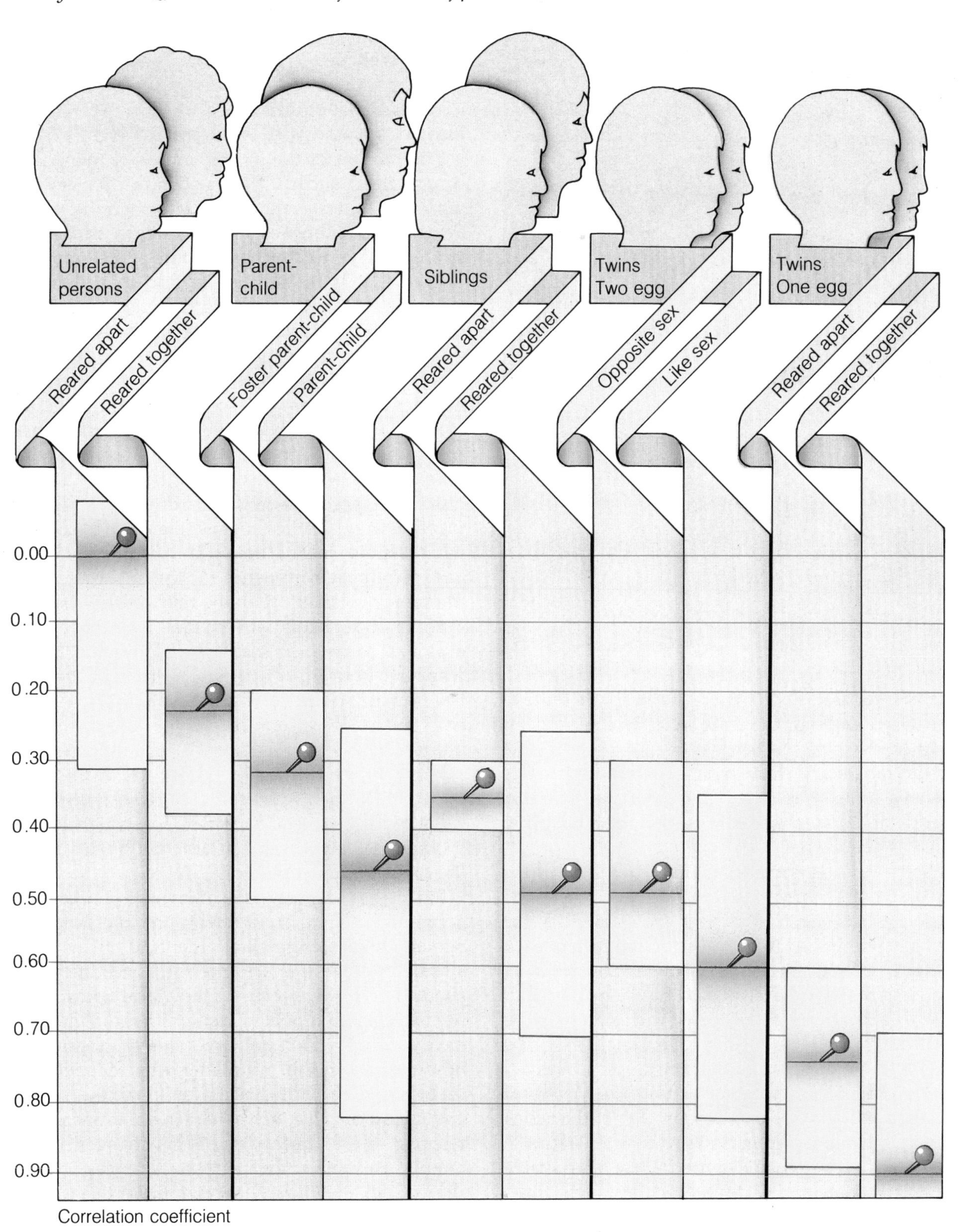

Once joined at the hip, this pair of Siamese twins have been separated by surgery. The tendency to give birth to twins is a hereditary trait, with an overall probability in the general population of 1 in 96 births.

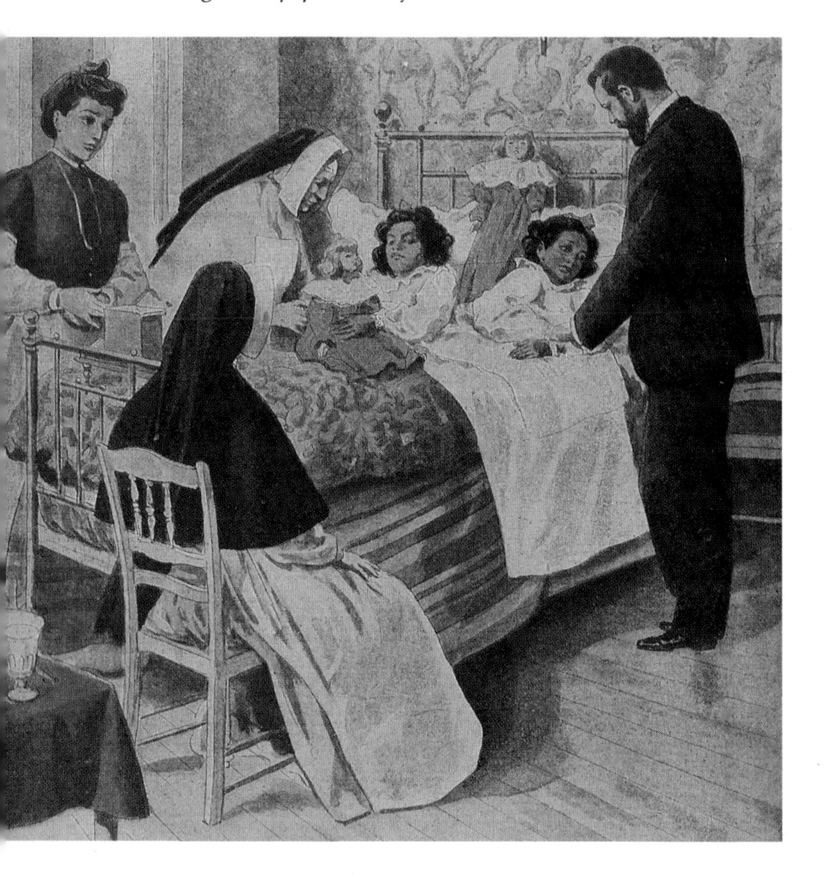

scrambled, as it is when the egg and sperm unite to create an embryo, then the exact mix needed for that trait is likely to be destroyed — what results is the hereditary characteristic as opposed to the genetic one. The trait may not be strongly hereditary, and is therefore passed on to offspring only as a rare event.

Some features such as height and weight can be assessed in a more quantitative way. Identical twins, whether raised together or separately, are usually within half to two-thirds of an inch of each other in height. Nonidentical twins and brothers and sisters are on average more than one-and-a-half inches apart in height. So the conclusion must be that adult height is mostly determined by genes and little affected by surroundings.

Weight is a more variable feature. The weights of adult identical twins parted soon after birth show a much greater difference than those of identical twins raised in the same home. Where it is possible to assess the social differences between these environments, such as by comparing income levels, then nearly always the heavier twin has come from the more advantageous background. It is perhaps not surprising to discover that higher incomes are associated with greater weight.

Comparison of the IQ test scores of twins is a controversial subject. No one knows for certain what these tests measure, but there is good evidence that the way an individual performs is strongly influenced by both culture and education. Therefore it is no surprise to find that identical twins raised together are more alike in respect of IQ than fraternal twins or brothers and sisters.

Identical twins living apart vary more and their scores resemble those of brothers and sisters or nonidentical twins. In some studies, pairs of identical twins have been compared for both IQ test score and educational attainment. Results show that the test scores become further apart as the difference in education standards widens. Taken together these studies suggest that IQ test scores, whatever they mean, contain both an inherited and an environmental influence and that each has about the same overall effect on the final result.

Some aspects of twin studies are difficult to evaluate. Many examples exist of identical twins who have been raised apart from birth that go on to have similar wives, life histories, children's names, pets and color preferences. These findings seem impressive evidence for a great gene influence on personality and lifestyle. However, it is very difficult to know what to compare these twins with. What part does coincidence play? How often would brothers or sisters raised in different homes show the same or similar patterns?

Finally twin studies have often neglected to take account of an important environmental factor — namely interuterine life. Identical twins are attached to the same placenta and hence the same circulation in the womb. Nonidentical twins have separate blood supplies. Identical twins therefore have a virtually identical environment for the first nine months of life, no matter what happens to them later. Recent evidence shows that toward the end of uterine life babies can hear sounds and respond to the effect of emotional changes in the mother, so some of their early experiences may be shared. No one knows how significant this is.

Severo Ochoa

Enzyme Engineer

Dogged persistence in research over many years can lead to nothing but frustration and disappointment. For Severo Ochoa, however, it led—eventually—to the award of a Nobel Prize, and to being recorded in history as the first to find evidence that metabolic energy is stored in or utilized by the body through high-energy phosphate compounds. Later he made the equally important discovery of a means to synthesize nucleotide chains parallel in composition to RNA.

Ochoa was born in Luarca, on the northern coast of Spain, in September 1905, the youngest son of a lawyer. Although he graduated from the University of Málaga with a BA in 1921, it was not until 1929 that he received his medical degree at the University of Madrid. For the next two years he studied at the Kaiser Wilhelm Institute in Berlin, after which he spent another twelve months in London at the Institute of Medical Research. In 1934 he returned to his native land to take up a position as a teacher at Madrid University, only to find that within two years more the Civil War disrupted his research. He spent the years 1936 and 1937 at Heidelberg, and 1938 to 1940 at Oxford. It was in the latter year that he first went to the United

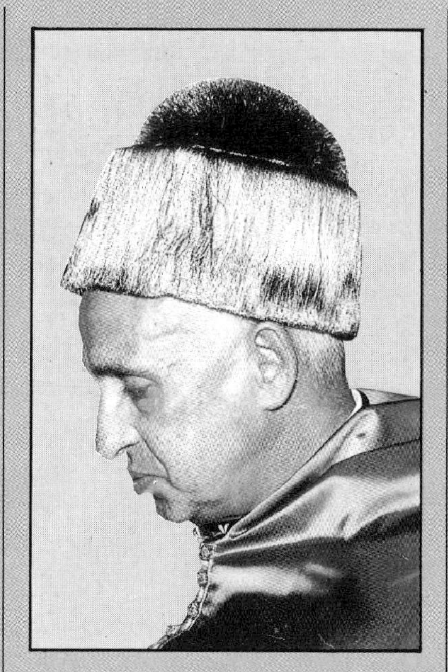

States, joining Washington University in St Louis as instructor and research assistant in pharmacology. Two years later he transferred to the faculty of New York University College of Medicine, with which he was associated for the rest of his life, retiring as Professor of Biochemistry.

At Heidelberg, Ochoa studied under the biochemist Otto Meyerhoff. His specialty was the physiology of muscle, and it was upon this basis that his later insights into the nature of how metabolism stores energy were grounded.

By adducing evidence that a cell stores or utilizes energy through energy-rich phosphate compounds, compounds such as ATP (adenosine triphosphate) and guanine triphosphate, that can be used as necessary, and by investigating the liberation or incorporation of carbon monoxide or oxygen, Ochoa discovered the enzyme polynucleotide phosphorylase.

It was already known that the nucleic acids RNA and DNA were made up of four different nucleotides, and that the body somehow combined the nucleotides to make the acids; an enzyme or enzymes was obviously involved. Ochoa had now found one.

Having isolated the enzyme, Ochoa went on to combine it with nucleotides to which a phosphate unit had been added, in order to try to produce a nucleic acid. Sure enough, long thin molecules of RNA were formed—but the process also revealed that any one of the four nucleotides could be produced singly.

(It is now known, however, that polynucleotide phosphorylase is in fact not the enzyme that normally assists in RNA synthesis in the cell.)

Ochoa's discovery was made in 1955. In 1956 the American biochemist Arthur Kornberg extended Ochoa's method and derived a synthesized form of DNA. In that year, too, both Ochoa and Kornberg received the Nobel Prize for their work.

A child's potential for physical and mental development is inherent in his or her genetic make up. But this potential can be maximized by a favorable environment, such as the best food and a good education.

A problem for students of nature versus nurture is raised by recent changes in the physical height of human races. Twin studies indicate that height is largely due to genes — but then how can the average height of Japanese men and women have increased so rapidly in this century, and is it significant that this occurred at the same time as Western food became popular in the Far East? Why are Californians, drawn from the same genetic pool as other Americans, significantly taller than the rest of the nation?

The explanation appears to lie in the effect of environment on gene expression. Although the genes available do not change — they can alter only over many generations — factors such as climate and diet have allowed these genes to more fully exert their influences. It seems reasonable that a very marked improvement in diet might permit greater growth. Presumably the twin studies did not include pairs with extreme differences in surroundings so that the environmental effect was not detected. For any given collection of genes the overall effect in a whole population is affected by socio-economic factors.

The ultimate interaction between a person's genetic constitution and his or her environment comes in the plastic surgeon's office. Here any degree of inherited ugliness can be modified, perhaps even reversed. While the underlying genes are not modified there will be no change in future generations, although changes to the characteristics that nature imposed can be dramatic. Surgery may even enhance the chances of passing on the set of genes a person was born with in that they can make it more likely that he or she will have offspring.

Recently a striking example of this interaction has become available. Children who are born with the genetic disorder Down's syndrome (mongolism) have a very characteristic facial appearance as well

People such as Eskimos (below) *have become genetically adapted to cope with the environmental conditions in which they live. The evolution of a metabolic response for body heat maintenance has enabled them to withstand conditions of extreme cold. An almost immediate adaption to the environment can be obtained by those who, shunned because of an ugly nose* (bottom), *undergo plastic surgery.*

as a moderate mental retardation. In particular they have folds of skin at the corners of their eyes. The degree of mental retardation varies greatly from person to person and children raised at home seem to attain a higher educational standard than those raised in hospitals.

Down's syndrome children are often shunned by other children because of their unusual appearance. This undoubtedly affects their mental and social development and may contribute to the severity of their mental retardation. The children's eyes can be changed by cosmetic surgery so that they look normal. Down's children who have had this done seem to be able to form more normal social relationships with their peers. This in turn could modify the extent of their mental handicap. Here, to add to the controversy about nature versus nurture, is a fascinating and valuable environmental modification of the phenotypic outcome of a genetic disorder.

Chapter 5

The Gene Machine

Whether you are a farmer in France or a housewife in Kalamazoo, genetic engineering will change your life: it is highly likely that you will be — if you have not been already — affected by the biotechnology revolution.

Strictly speaking, genetic engineering is only one component of the explosion in the biological sciences. It is closely related to advances in cell biology, test-tube fertilization and molecular biology, which together promise to alter our world. What is happening is both simple and dramatic. It is now possible to transfer — or even to manufacture — genes artificially and to make them perform their normal functions. New types of plants and microorganisms can be created almost at will. A single desirable property can be transplanted into a single cell and then expanded many billionfold. Within a few years, everything from the food we eat to the medicines we take will probably have been modified by genetic technology.

The Keys to the Kingdom

Genes are made of DNA (deoxyribonucleic acid), which normally occurs in the form of long strings. When scientists extract it from a cell, it has a marked tendency to form sticky, gluelike blobs. The strings get hopelessly entangled, like the noodles in a plate of spaghetti. To make matters worse, the molecule is composed of a nearly endless string of bases with no obvious starting point at which a scientist could attempt to break it down into regular fragments.

Genetic engineering began with the discovery of special proteins called restriction enzymes, which have the ability to slice DNA up into regular, reproducible pieces of a "manageable" size.

Restriction enzymes are the pattern-cutting tool of the genetic engineer. The first useful ones were discovered in 1970 by Hamilton Smith at Johns Hopkins University in Baltimore. There are now hundreds of different restriction enzymes available to scientists, each capable of chopping up a piece of

The marvelous feat of engineering that is a microchip less than one-tenth of an inch across contains a staggering amount of information in its tiny size, which allows it to decode and transmit messages. In the same way DNA and the genes control the factors of inheritance in their microscopic area. Scientists have now learned to engineer these biological storehouses so that they can manipulate and control the formation of the end product —a living organism.

The bacterium Escherichia coli, *here freeze-fractured and stained, is used in genetic engineering. It contains enzymes able to cut fragments from DNA. The fragments can be added to other DNA pieces.*

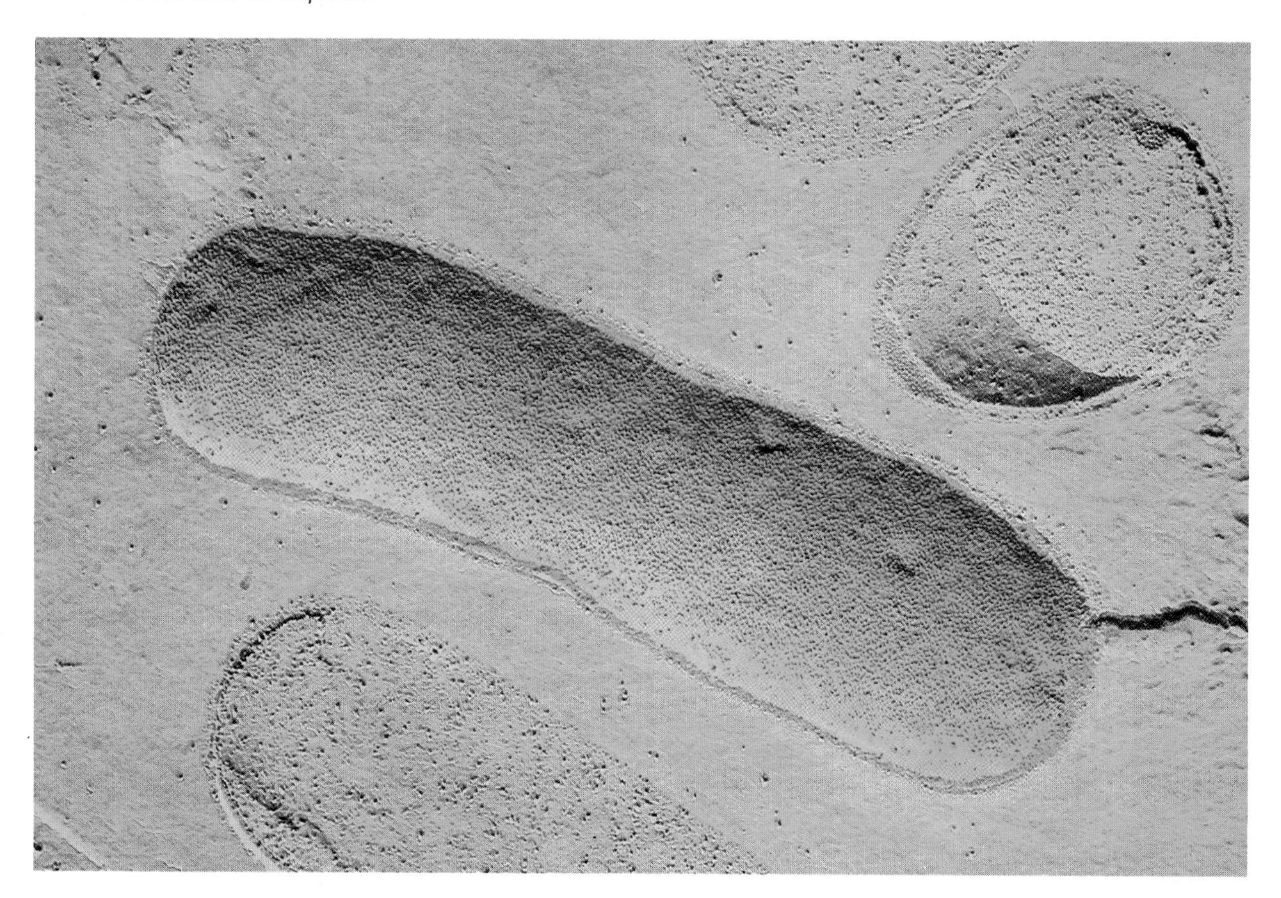

DNA in a slightly different way, enabling each fragment produced to be analyzed. The sequence that controls life has become visible.

How do these useful molecules work? Almost all restriction enzymes, isolated from bacteria, cut a DNA strand only when an exact recognition sequence of bases occurs. For example, the enzyme EcoR1 is isolated from the bacterium *Escherichia coli* RY13. It binds to and snips a DNA strand only if the exact mirror-image dual sequence $\frac{\mathrm{G}\downarrow\mathrm{AATTC}}{\mathrm{CTTAA}\uparrow\mathrm{G}}$ is present. The molecule is cut where the arrows indicate and nowhere else. It is a powerful tool, and is capable also of revealing the exact sequence of bases at the ends of the pieces. If the EcoR1 does not break the DNA into a piece with the sequence sought experimentally, that particular sequence of bases is not present. This is quite an amount of information from a single "snip."

When the effects of different restriction enzymes are compared, each restricted to different base sequences, it is possible to break down a large piece of DNA into several smaller, more manageable fragments. Certain restriction enzymes always chop up one specific piece of DNA into the same-sized fragments. This is an important property and is the basis of one type of genetic analysis of great value to genetic counselors. It allows human carriers of some abnormal genes, who are not themselves affected, to be identified and advised.

Once regular-sized chunks of DNA could be reliably produced, it became possible to use them to determine exactly how genes were constructed and how their activity might be controlled. An important development was the subsequent ability to produce large amounts of exactly the same piece of genetic material. This is done by stitching the fragment needed into a plasmid, a type of extra gene structure on a bacterial chromosome.

Plasmids normally exist as tiny circles of DNA. They can be snipped open with a restriction

Daniel Nathans and Hamilton Smith

Cutting Chromosomes

For some time after the composition of DNA had become known, scientists remained baffled as to the actual ordering of genes on a chromosome, or even as to the specific chemical constituents of the genes themselves. Then, during the 1960s, Werner Arber — a Swiss microbiologist — discovered restriction enzymes. These snap apart (or cleave) the chain of genes at distinct locations on the DNA molecule. By using a combination of such enzymes, individual genes or gene fragments (very short chains) can be separated and analyzed. It was at this point that Daniel Nathans and Hamilton Othanel Smith made their contribution.

Nathans was born in October 1928 in Wilmington, Delaware. He graduated from the University of Delaware in 1950, gaining his medical degree at Washington University, St Louis, in 1954. The next eight years were spent first at the National Cancer Institute, then the Columbia Presbyterian Medical Center, and subsequently the Rockefeller University in New York. In 1962 he became Professor of Microbiology at Johns Hopkins University, Baltimore.

Smith was born in New York City in August 1931. Much of his final education was at the University of Illinois, but it was

from the University of California at Berkeley that he graduated, and — unusually for a molecular biologist — it was in mathematics that he did so. His medical degree was awarded him by Johns Hopkins University in 1956, following which he has devoted himself almost entirely to research there.

It was while working with the organism *Escherichia coli*—a bacterium usually present in the human gut — that Arber first discovered restriction enzymes. Smith later and quite independently worked on the bacterium *Hemophilum influenzae*, and managed to isolate an enzyme (later called Hind II) which also cut DNA at specific sites. He thus confirmed Arber's findings (and others have since gone on to isolate a large catalog of similar enzymes), but was the first to be able to identify the gene fragments thus produced, opening up the field to the identification of individual genes within the chromosomes.

Since his time at the National Cancer Institute, Nathans had long been interested in cancer research. Now, working in part also independently, he studied the carcinogenic virus SV40. In 1971 he showed that it could be cleaved into eleven identifiable fragments, thus also demonstrating that genes could at last be fully and accurately mapped, from which it also followed that DNA might someday be synthesized. Another spinoff was the great increase in the capacity for genetic engineering.

In recognition of their original research, Nathans and Smith shared the 1978 Nobel Prize for physiology or medicine, together with Werner Arber.

The recombinant DNA technique can be used to produce human proteins such as insulin (below). *Plasmid DNA is removed from a bacterium, as is DNA from a human cell. Restriction enzymes then cut the plasmid open and cut the insulin-producing gene from the human DNA. This gene is then inserted into the plasmid, which is put back into the bacterium. Several bacteria similarly engineered are multiplied in a fermenter. The insulin they produce is separated from them and purified for use as an injection for diabetics. The DNA of the lambda bacteriophage* (bottom left) *is used to clone genes from other organisms.*

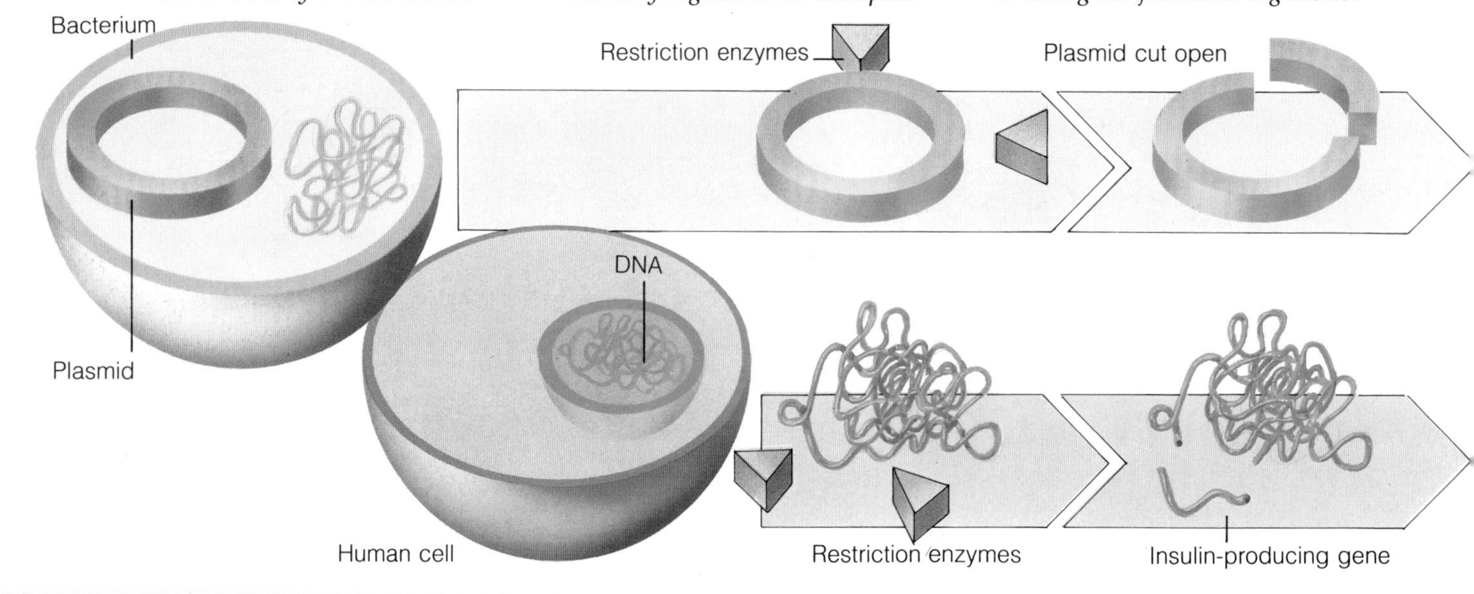

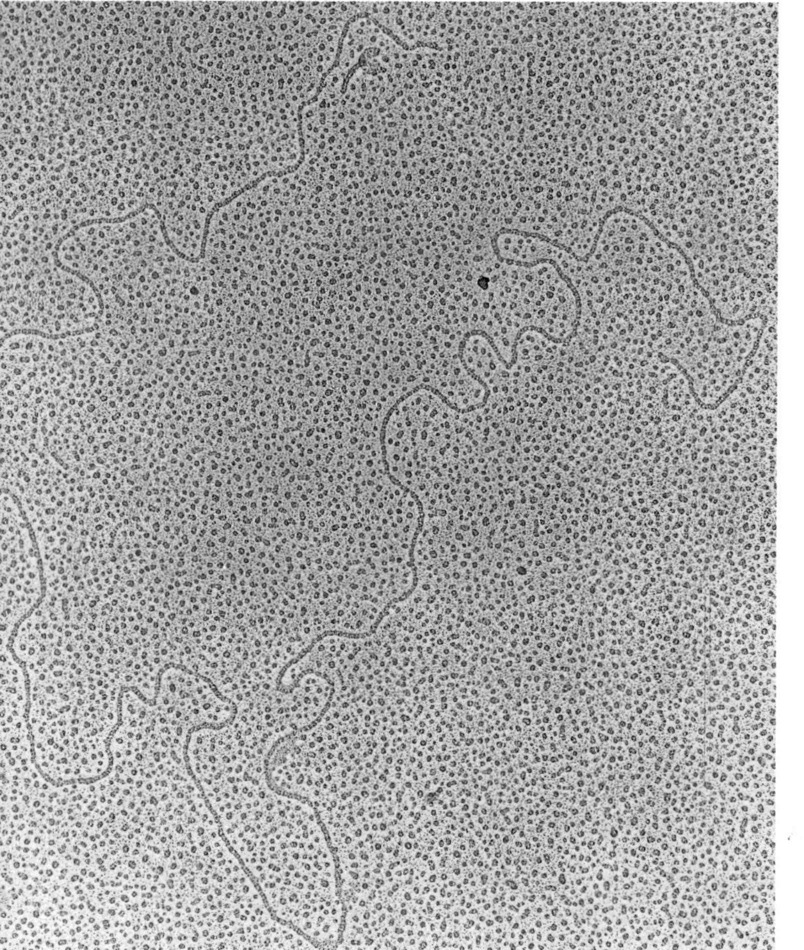

enzyme and then reformed with the extra fragment inserted into the gap. (This is possible because the DNA cut by the enzyme has "sticky ends" that bind the new fragment.) The plasmid, with its new piece of DNA, can then be reinserted into its normal bacterial host. If the bacteria are grown in large quantities, the plasmid can be recovered and the inserted fragment purified. The process is called gene cloning. A single gene is isolated, inserted into a carrier microorganism, grown, and clones — billions of them, all identical copies — are produced. The technique is often known as recombinant DNA research, because it involves the recombination of several pieces of DNA.

The first genes were cloned by Herbert Boyer and Stanley Cohen at Stanford University and the University of California at San Francisco, in 1973. The explosion of genetic experimentation was about to take place.

Public Relations

When news of the early gene cloning experiments reached the public there was a near-hysterical reaction. Why was recombinant DNA so worrying?

One horror story, widely circulated, was that scientists would soon clone cancer-causing genes into bacteria. If these recombinant organisms escaped from the laboratory, the whole world might suddenly become infected and die. This was never a real possibility. First, the gene would be

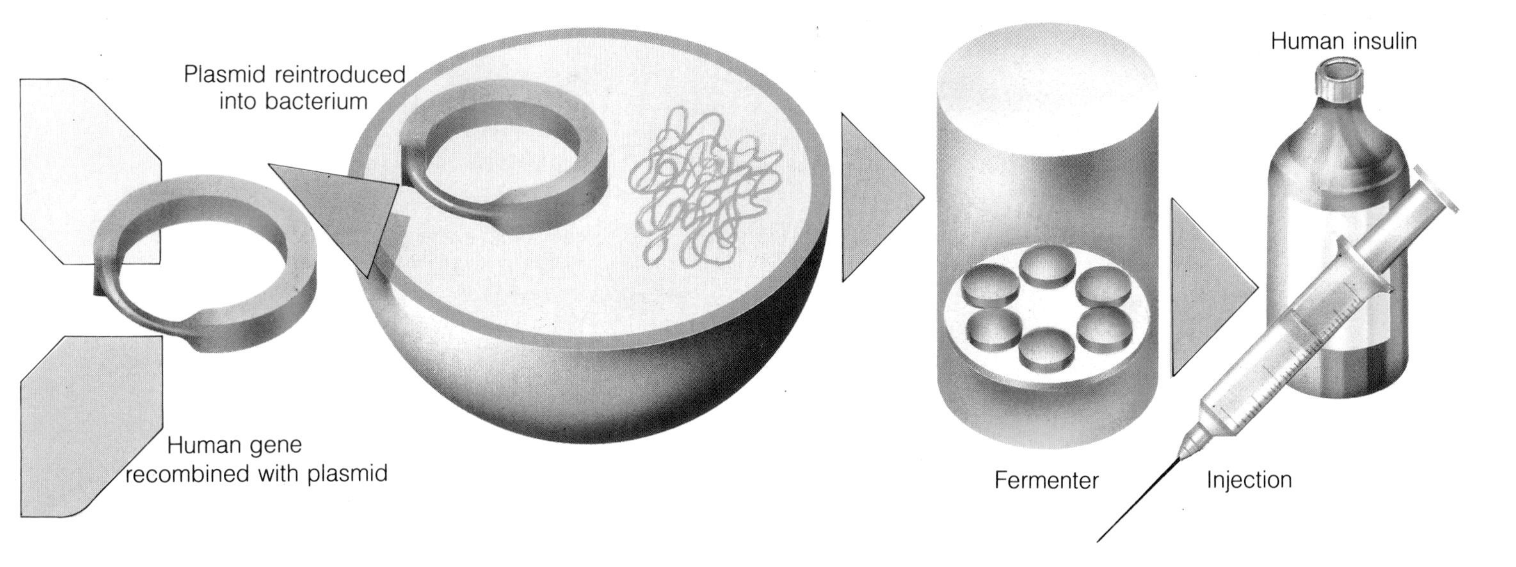

locked up inside the plasmid inside bacteria and should never come into contact with a human cell. Second, the special strains of *E. coli* that were used were so fragile that they were very unlikely to survive when confronted by their robust cousins in the human gastrointestinal tract. Finally, genes need to be activated to produce their effects and, in the early days at least, none of the signals that controlled genes had been identified.

Another somewhat more feasible possibility was that an unscrupulous scientist or terrorist group might develop "super bacteria" designed to contain the information needed to produce multiple toxins — such as tetanus toxin — but resistant to all known antibiotics. If such a bug were released it would wipe out humanity overnight. The threat of this alone might be used as a basis for international blackmail.

Working with microorganisms is a little like working with radioactive materials. Sophisticated facilities are needed to protect both the experimenter and the experiment from contamination. If even a single ordinary bacterium got among the delicate specimens needed for genetic recombinant work, it would rapidly affect the whole culture.

Despite these inherent safeguards, the "mad scientist" theory caused considerable concern. Once politicians, the press and a small number of scientists had cried "Wolf!" it became a matter of great public interest. People wanted to know exactly how dangerous genetic manipulation was, and asked if it should be stopped. The mayor of Cambridge, Massachusetts, home of both Harvard and the Massachusetts Institute of Technology, banned all such experiments from the city. Other areas threatened to follow suit. Congress was called upon to enact legislation prohibiting or strictly controlling all experiments involving recombinant DNA. In order to try to head off a complete moratorium on all genetic engineering, a group of scientists involved in this research gathered at Asilomar in California during 1975 to draw up a set of guidelines.

One of their recommendations was that only specific genetically crippled organisms should be used as carriers, making the possibility of their successful escape from the laboratory impossible. Such organisms were created in such a way that life for them was confined to the inside of a test tube. If an accident did occur, then there was no possibility that disease or environmental disruption would follow.

As usual in the United States the next stage was a review of procedures by a committee set up by the National Institutes of Health (NIH). Since NIH effectively controls the money available for funding recombinant DNA work, its decisions have the effect of law within the scientific community. Because NIH receives the money it dispenses from Congress, it is exceptionally sensitive to public

Scientists carrying out research in which genes in bacteria are cloned have to conduct their experiments in enclosed, strictly sterilized surroundings. The strains of bacteria used are so fragile that most could not survive outside their protected environment. For these reasons, it is highly unlikely that any cloned microorganisms could "escape."

pressure exerted through members of Congress. The Asilomar guidelines were further tightened, making work on potential cancer-causing genes as good as impossible.

The effect of all this on the scientists involved was that of unparalleled anticlimax. One day, a breakthrough into a whole new realm of experimentation and possibly the realization of the wildest dreams of molecular biologists — a complete understanding of how genes worked. The next day, a battle to retain the ability to perform only a peripheral few of the limitless experiments. There was even sufficient public hysteria to make it seem likely that these few strictly regulated studies might be condemned.

In time the storm was weathered, partly because the scientific community managed to reassure the public that they would perform their experiments with the utmost responsibility, and partly because of the fear that if work were banned the rest of the world might reap all the benefits of the American investment in genetic research.

It was eventually decided that as long as each experiment was approved by watchdog committees, and the safety facilities of any institution rigorously scrutinized, DNA recombinant work could continue.

Since 1976 general reassurance as to the safety of genetic manipulation has become widespread. It has been possible gradually to drop most of the restrictions. In 1979 work on cancer genes was finally permitted — and the result was a sudden new insight into the origins of human cancer.

Although the public seems satisfied with the safety of genetic manipulation experiments, a small group of activists has continued to fight to prevent some of the applications of these studies. Using the courts, these groups, led by Paul Rifkind, have sought to prevent the release of any genetically altered organism into the environment. Their argument is that it is not possible to completely predict the full environmental consequences of release and therefore it should be prohibited.

One particular case in point is a bacterium that has been slightly modified so that the temperature at which ice forms around the organism is lowered by about two degrees. The normal counterpart of this recombinant bacterium is responsible for the

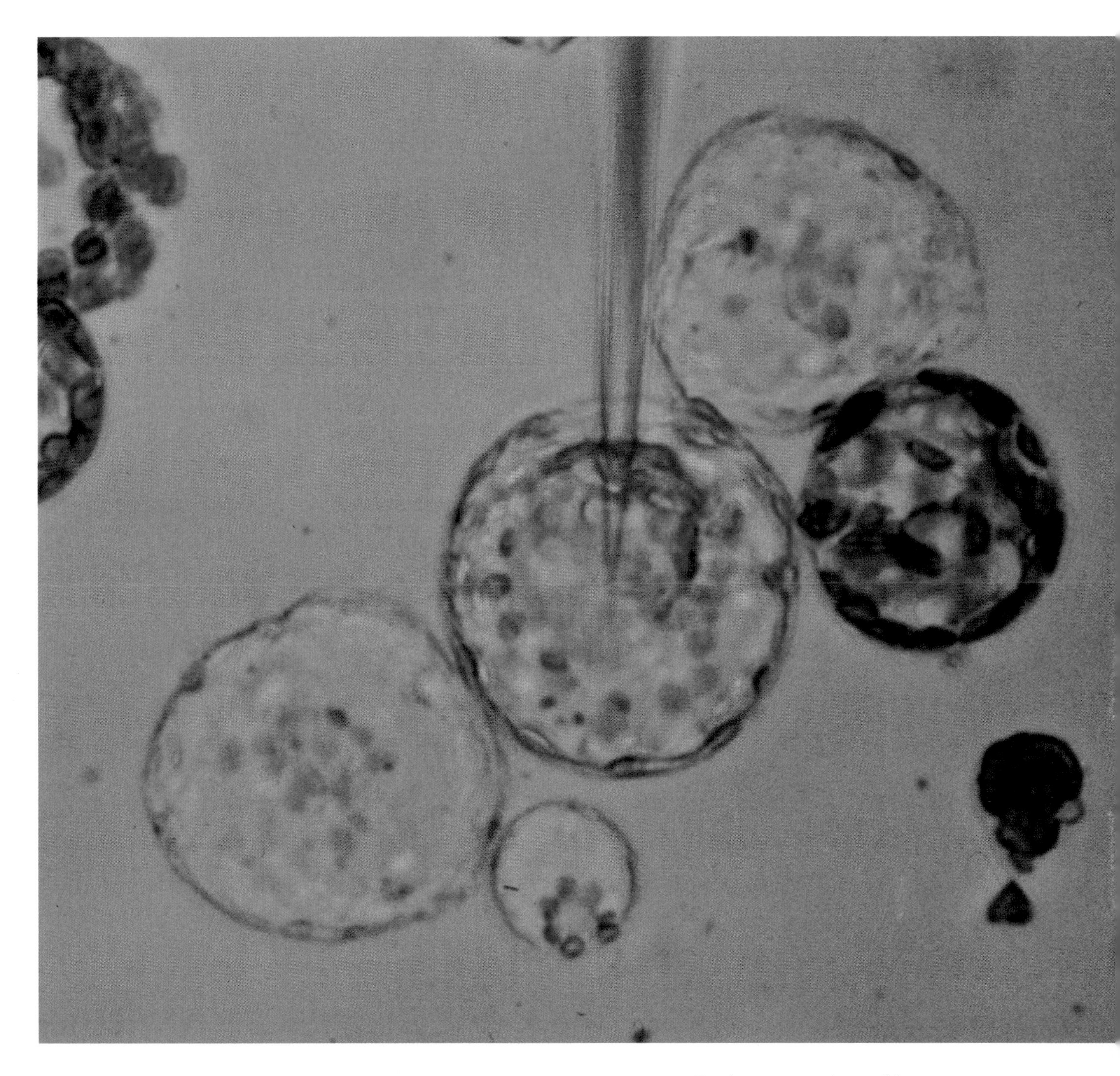

Foreign genes are inserted into a nucleus with a micropipette in a method called microinjection. The DNA substance here is dyed yellow so that it can be seen to pass exactly into the nucleus.

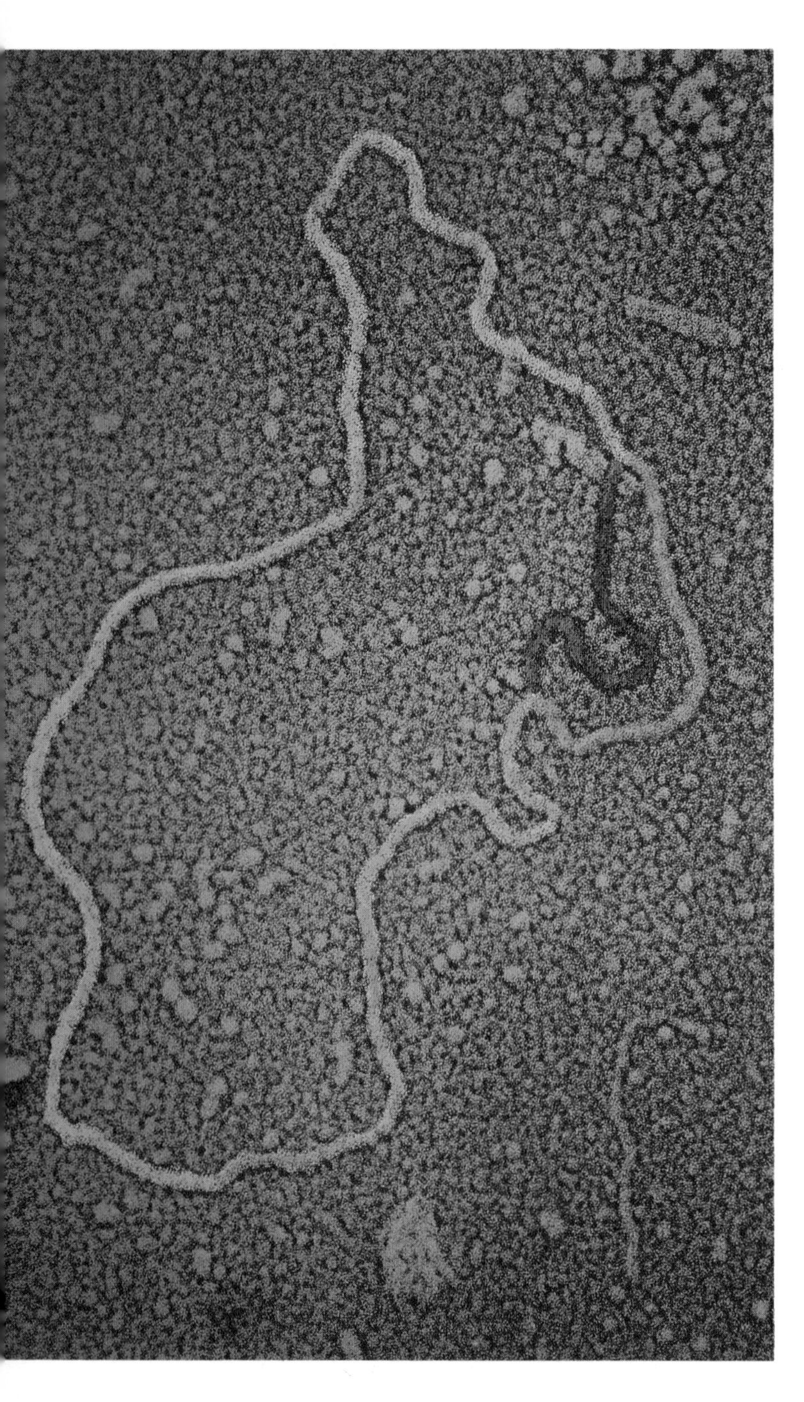

An electron micrograph of a bacterial DNA shows plasmid DNA (the blue and red loop), often used in genetic engineering experiments. The blue section is an individual gene, and the yellow is ordinary bacterial DNA.

The frost that fringes these bearberry leaves also affects commercial crops. Recombinant DNA could alter certain bacteria on the leaves so that they raise the temperature around them and reduce frost damage.

deposition of ice on the leaves of some commercial crops, such as lettuce, during a cold snap. It is thought, therefore, that frost damage might be prevented if the new organism replaced the old one. To commercial growers the economic value would be enormous because frost would damage crops less often. At least one case of this kind is still before the US courts and the eventual outcome of such actions may influence the course of genetic engineering in the future.

What Can be Learned?

Recombinant DNA technology puts the entire stock of biological information in the hands of scientists. Everything living has its basic plan encapsulated in its DNA. Genetic engineering is like discovering the keys to the ultimate library — once inside it is possible to choose and read any book. This is because DNA and the genetic code are universal; all living organisms use the same chemistry and the same code to store their blueprints. The tools developed to study primitive bacteria and viruses can be used with only a little modification to study human genes. Simple tricks can be played on primitive organisms such as yeasts and bacteria to get them to make human proteins. Genes from one species of plant can be cut out, "pasted" into a carrier and slipped into another species, so combining two desirable features.

So many different genes have been isolated that it is difficult to decide which is the most important guide to the uses of genetic recombinant technology. Perhaps the most exciting story is the discovery of oncogenes — specific genes thought to be involved in the development of cancer.

For more than a century scientists have speculated on the fundamental mechanism that changes normal healthy cells into dangerous enemies which can kill their host. The first clues were the discovery of groups of viruses that could cause cancer in experimental animals. Viruses have very simple genetic material, making it possible to obtain large amounts of it. This pure DNA could be used to make tissue cultures of mouse cells malignant — they were transformed. Chopping up DNA with restriction enzymes led to the identification of a specific gene-sized piece that contained all the transforming information present

In experiments to isolate genes it has been discovered that particular genes — oncogenes — produce cancer, or make cells malignant. The outgrowth in the center of the picture below is an example. The tools used to discover these genes are cells such as mammalian cells (bottom) *and yeast cells* (right). *When modified by recombinant DNA, these cells produce genes and proteins, which can be used to study human genetics.*

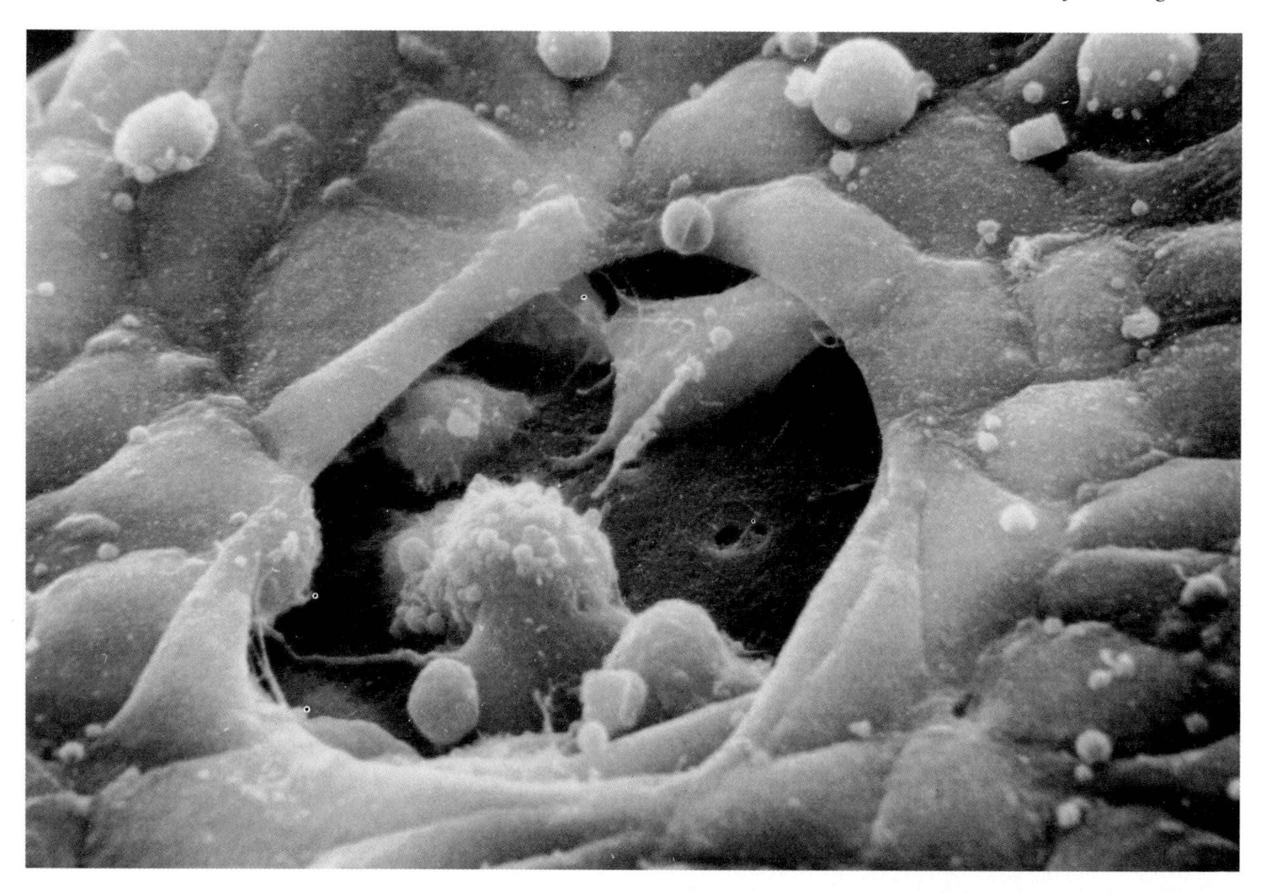

in the whole virus. A single gene was found to be responsible. Such genes are known as oncogenes (cancer genes), and about thirty of them have been discovered so far.

Perhaps the most far-reaching discovery was that oncogenes are not really virus genes but are normally present in all cells. The virus only picks them up and moves them to a new, abnormal environment where they somehow make recipient cells malignant. Initially this discovery was a great shock. How could apparently normal genes produce cancer? And does this mean that cancer cannot be controlled?

The normal counterparts of oncogenes seem to be involved in controlling the growth of normal cells. Like instruments in an orchestra they each have a piece to play in the coordinated symphony of normal regulated cell division. When damaged, or moved to the wrong place, the delicate controls that direct exactly when each gene should perform are

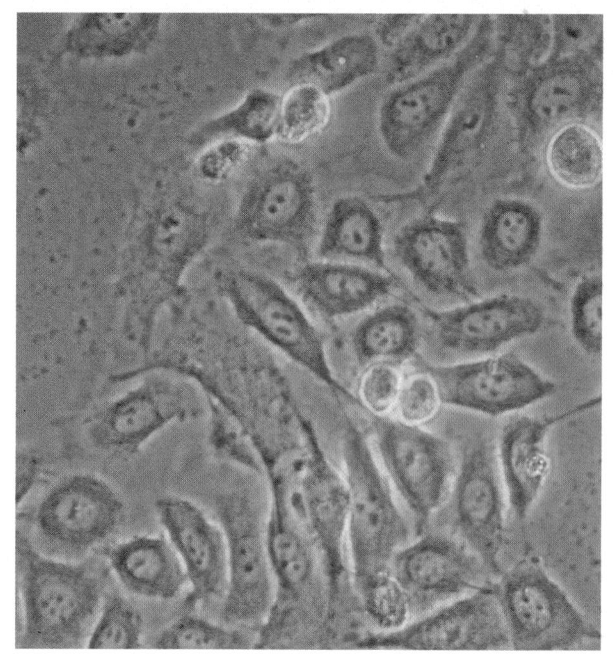

An orchestra relies on coordination to produce a harmonious piece of music, just as normal cell growth depends on the organized workings of genes. Dysfunction in both cases results in haphazard disorder.

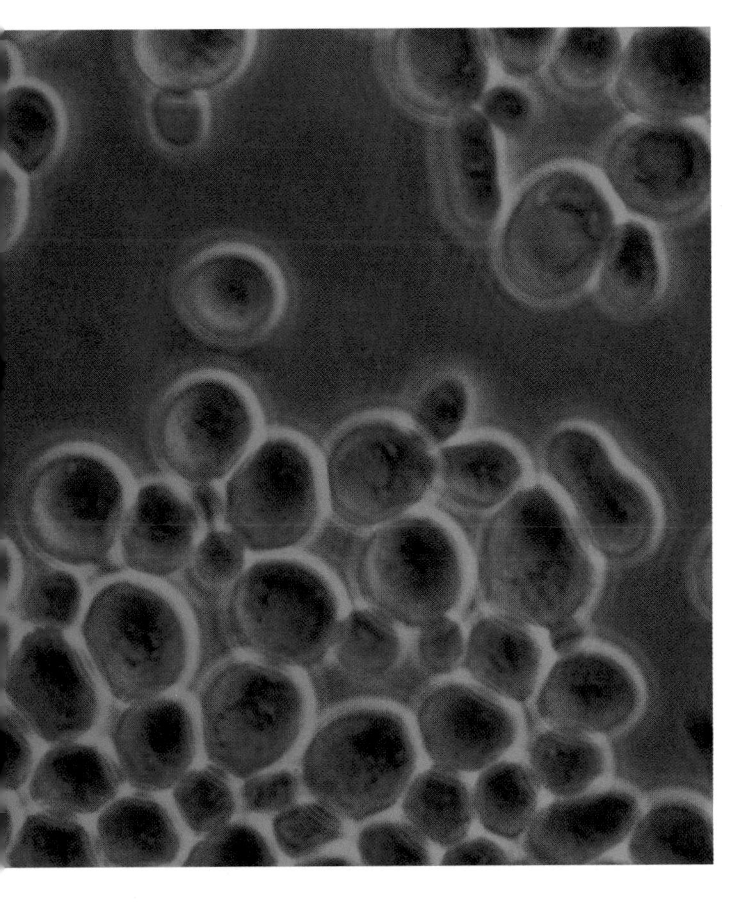

lost. Like an orchestra without a conductor, the coordination between instruments disappears.

Although the exact normal function of many oncogenes is not yet known, it is encouraging that fewer than a hundred genes out of a possible million are involved. This gives scientists a smaller target to aim at. Already a whole new branch of anticancer drugs designed to reverse or inhibit the effects of oncogenes is predicted.

Detecting Abnormal Genes

A major goal of genetic engineering research is to identify abnormal human genes and reverse their effect, or at least give parents a choice of whether or not they wish to allow defective children to be born.

The important property of DNA that allows this type of genetic research is the existence of two mirror image strands which stick tightly to each other. When DNA is heated it "melts" so that the two strands come apart. When it is cooled, the complementary halves find each other and re-form their partnership. If single strands of DNA similar or identical to the "melted" DNA are added, then a strand will stick onto its complementary piece when the strands reassociate. If this small piece of DNA (called a probe) is labeled with radioactivity, it is possible to detect whether it has combined with the test sample. Probes can be used to identify particular genes or even chromosomes.

As has been seen earlier, a particular restriction enzyme always splits off a particular piece of DNA into characteristic fragments. It is therefore possible to chop up DNA and then separate the pieces by size. When a labeled probe is added, it sticks only to the complementary piece of DNA, and the size of the DNA fragment is known. If everyone had identical DNA, then the fragments would be exactly the same size. But most of us differ in apparently insignificant ways, so that the sizes of fragments that bind a particular probe may be

Mouse cells are often used in experiments to test the receptiveness of mammals to cancer-causing agents and to detemine whether the cells can be engineered to reduce the negative effects of such agents.

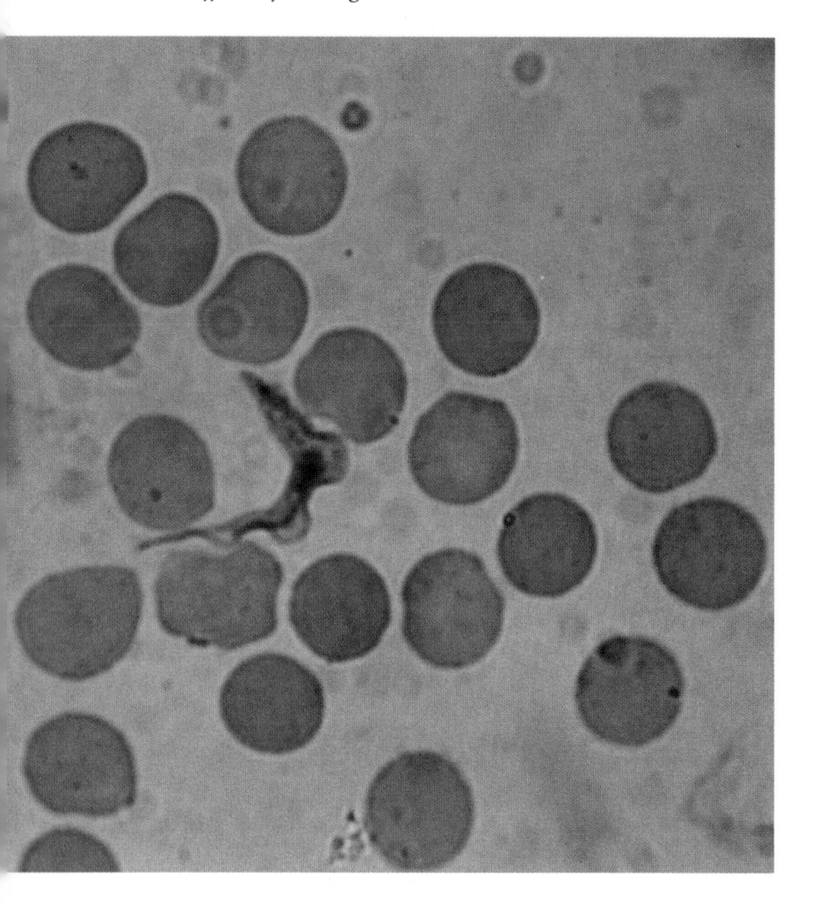

different from person to person. This is called a restriction fragment length polymorphism (RFLP).

RFLPs are useful if they occur near a gene that is to be studied. They provide markers for a particular stretch of DNA and allow scientists to determine who has, and who has not, got that marker. In family studies these markers may be used to follow the inheritance of a particular gene.

Of course, it takes many painstaking studies of many large families to work out the relationship of the polymorphic marker to the disease, but once an RFLP has its value established then it becomes a useful tool. The particular marker need not be the defective gene itself, as long as it is close enough for the RFLP and the gene to be inherited together. In an ideal world, the probe would always land directly on the disordered gene and everyone who carried the gene would have exactly the same change in their DNA, so that the pattern would always reveal itself clearly as "defective." However, this is not true in many cases, and genetic counselors have to make do with less than perfect markers, but in other cases useful information may be obtained.

Two examples from clinical practice demonstrate how RFLPs work in the real world. Duchenne-type muscular dystrophy causes severe muscle wasting and early death in boys. The gene responsible is carried on the X chromosome, so that the mother has one normal and one abnormal chromosome. The mother of an affected child may go to the clinic for advice. Can she risk ever having a son, or should all her male fetuses be aborted (when only half of them may be affected)? Can she be assured of having an unaffected boy?

The geneticist can use RFLP markers to see whether he can detect her two different X chromosomes. This is not always possible, but if he can tell them apart, then he may be able to identify the defective one, which will also be found in the affected son. If the suspect chromosome can be identified for certain, then it is possible to tell very early in pregnancy whether a male embryo growing in the mother has received the good or damaged chromosome. A tiny fragment of placenta is safely removed before the tenth week of pregnancy and its DNA analyzed. If the disease-associated RFLP is missing, then the child should be healthy.

The same genetic analysis can be broadened to include all the members of an extended family. If a condition like muscular dystrophy is known to exist in the family, there is a risk that many of the daughters may be carriers. If the suspect chromosome can be identified, it may be possible to trace it through the family. In some cases it is possible to tell the sisters of an affected boy whether they are carriers or not. The chances of passing on the damaged chromosomes are fifty-fifty, and this must weigh heavily on the minds of potential carriers. Genetic marker analysis can convert this uncertainty to near certainty in some cases, allowing rational choices to be made. It is not difficult to imagine a person's joy at being released from worry that the harmful gene may be passed on to future generations even though the person himself or herself is normal.

Many RFLP markers for the X chromosome have become available. These enhance the accuracy of the analysis and also make it possible to find suitable markers for more families. In the near future the whole structure of the X chromosome may become known.

Gene Therapy

What are the prospects for replacing or masking defective genes? This subject is in its infancy because it requires getting a gene linked up with exactly the correct control signals, so that it could be switched on and off when required. But it may prove even harder to get the linked-up gene into the right cells where it is supposed to act.

There is one experiment, however, which shows potential. The growth-hormone gene was inserted into mouse embryo cells attached to a control switch activated by the metal zinc. Mice with the gene inserted and a high level of zinc in their diets grew much larger than their unengineered cousins. Although these experiments stimulated numerous comments about "supermice," it does show that in principle gene therapy might work.

An unauthorized attempt to perform gene therapy in humans was made a few years ago. It caused a furor. Two people suffering from fatal anemia were given their own bone marrow cells which had a normal hemoglobin gene inserted into them. It was hoped that the normal gene would function and allow them to produce normal, or at least better, red blood cells. These gene transfers did not work. NIH found out that its guidelines on recombinant DNA work had been breached, and took action against the scientist concerned.

The first approved trials of gene therapy are due to have started by early 1986. They will involve trying to insert the gene for a missing enzyme into the cell of an afflicted child in the hope that enough enzyme will be produced to relieve the symptoms. Such gene therapy — germ (sex) cell therapy as well as somatic cell therapy — may help the sufferer, but will not alter his or her DNA. The gene may still be passed on. If gene therapy does prove to be successful, very careful genetic counseling would have to follow.

Cloning Animals

Along with the revolution in molecular biology there has been a marked advance in the ability to

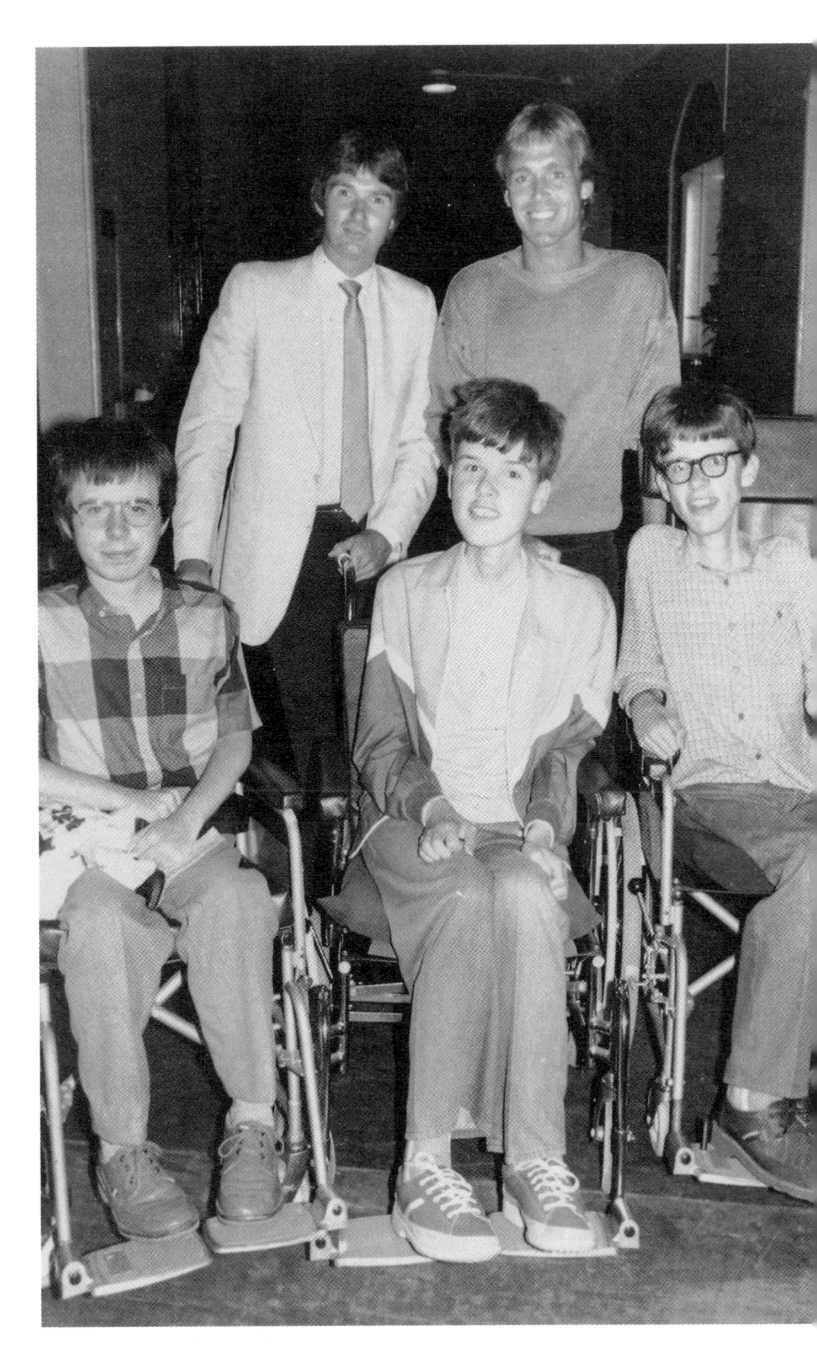

One type of muscular dystrophy, which affects mainly boys, is carried genetically and can be detected in a woman before she bears a child. She can then be advised whether or not to risk having an affected son.

The idea of cloning humans, or at least engineering them genetically to produce an ideal, captured the imagination of the Germans in the 1930s, who displayed posters of the Aryan ideal: blond and blue-eyed.

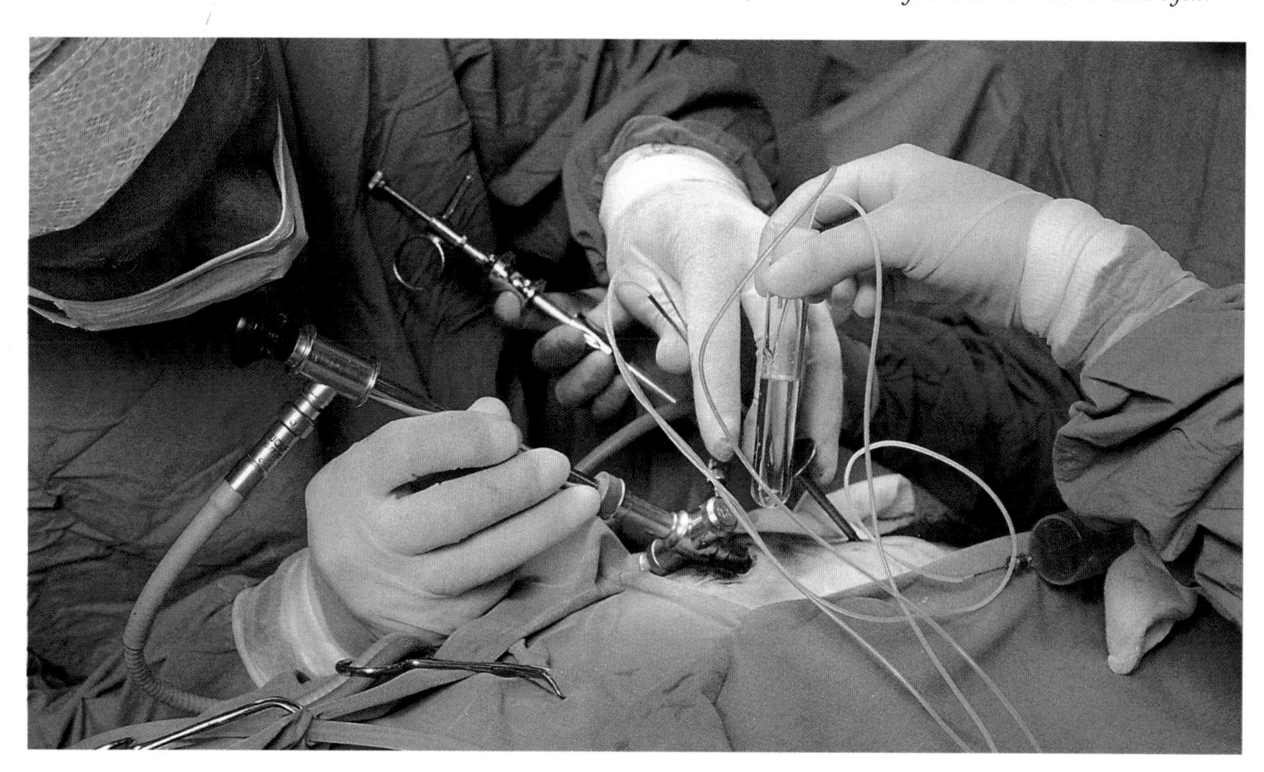

Many women cannot conceive because of malfunctioning ovaries or damaged Fallopian tubes, although they are able to nurture an embryo in the womb if it is placed there artificially. The Graafian follicle is drawn from a woman when she ovulates (above) *and the egg is separated from it using microinstruments* (right). *It is then fertilized by sperm on a tissue-culture plate and developed for a few days in a test tube, before it is implanted in her uterine wall.*

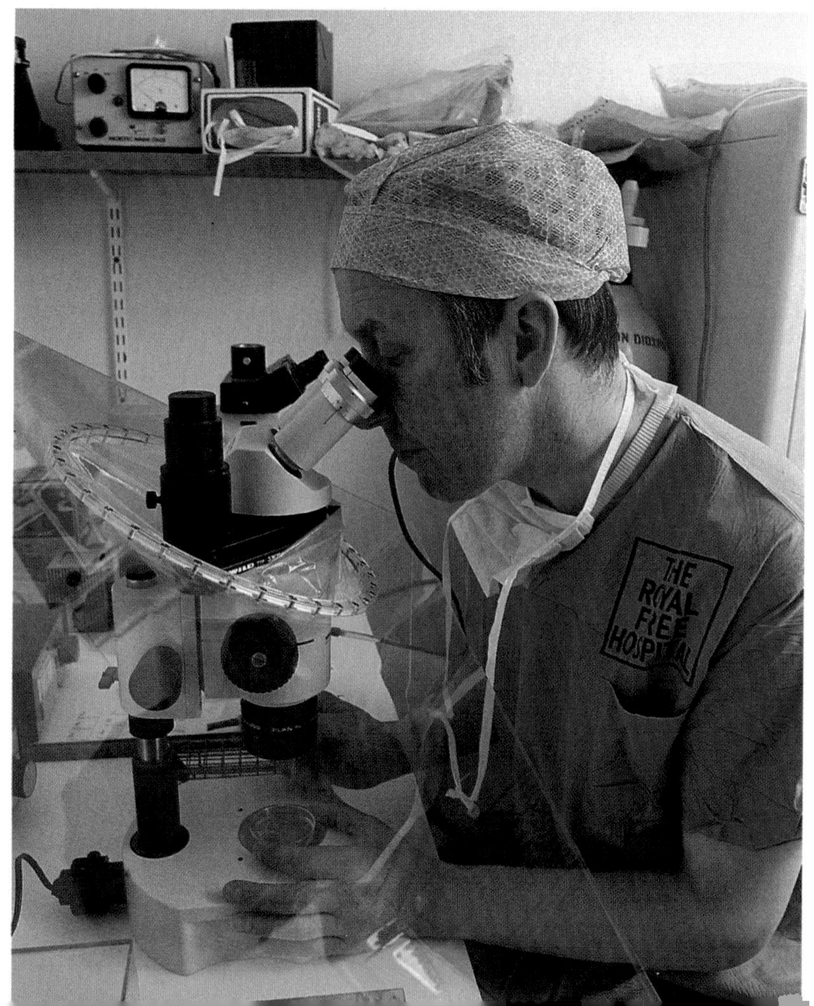

work with live cells. Possibly the most dramatic examples are test-tube babies and "clones" — identical copies of a single individual.

Test-tube babies are produced by fertilizing a woman's egg cell with her husband's sperm in the laboratory, and then implanting the fertilized egg into the woman's womb. For many women who suffer from damaged Fallopian tubes, in which eggs are normally transported from the ovary to the womb, there is no other way they can become pregnant. Careful hormone manipulation ensures that the eggs are produced at the right time. These are harvested, mixed with the husband's sperm in a tissue-culture dish, and then grown for a few days in the laboratory to mimic the time a fertilized egg takes to complete its journey down the tube to the womb. While this is happening the mother's womb is prepared to receive the embryo. If all goes well the minute embryo attaches itself to the wall of the womb and grows just like a normal embryo.

Until very recently it seemed impossible that the delicate egg and even more delicate tiny embryo could be handled or grown outside the body. Now it is almost commonplace and, like recombinant DNA experiments, a storm is brewing about the consequences of the procedure.

Many women release several eggs when stimulated by appropriate hormones and there are usually some fertilized eggs left over after four to six have been implanted in the womb. In many cases the extras are frozen in case the first attempt to procure pregnancy is a failure. In one recent case in Australia the parents of frozen embryos were killed in an airplane crash. Who was responsible for the potential children? If the cells were discarded was it murder? If implanted into one of the many volunteers would a child have a right to the original parents' estate?

However difficult the dilemma in this case, it is minor compared to the issue of embryo experimentation. Can the spare fertilized eggs be used for experiments? Scientists suspect that many genetic disorders affect very early embryos. This process cannot be studied except by experiments on test-tube embryos. The moral, religious and legal issues raised are staggering.

In veterinary practice test-tube fertilization promises to revolutionize animal breeding. Large

Frogs have been used in cloning experiments because they produce reasonably large eggs in vast quantities. The clones shown here are those of Xenopus, *a South African frog, crossed with a clawed toad.*

numbers of eggs from a particularly choice cow can be fertilized by best bull sperm and many offspring reared in surrogate cows. The old limit of one offspring per mating has been overcome. Frozen embryos can be shipped anywhere in the world for breeding, and the herds of many nations can be improved by importing embryos with desirable hereditary features for local foster mothering. The technique is much cheaper and faster than importing adult breeders.

Clones of identical animals have stimulated many imaginary stories of cloned humans. The difficulty is to remove the nucleus of an egg and replace it with the nucleus of another cell. Imagine how delicate and sharp the needle must be to puncture a cell membrane and suck out the nucleus without destroying the complex cytoplasm. Not surprisingly this experiment is easier to perform with frog's eggs, which are much larger than the eggs produced by female mammals.

In the original experiment the nucleus of a frog intestinal cell was transferred to an egg. Because the new nucleus had two complete sets of chromosomes, the egg behaved as if it had been fertilized and developed into a normal frog. The importance of this experiment was that it showed that all the genes are kept in every cell nucleus, and that they can be activated if the correct signals are received. Differentiation, the development of a special type of cell like an intestinal cell, does not involve the permanent loss of any genes.

Occasionally it is useful to have a group of genetically identical animals for a particular experiment. Some controversies about the relative roles of heredity and environment can be studied only in this way. But in general, cloning many copies of one animal by repeating the frog experiment with many eggs and nuclei from one individual is not a very productive experiment.

Cloning techniques are particularly suited to

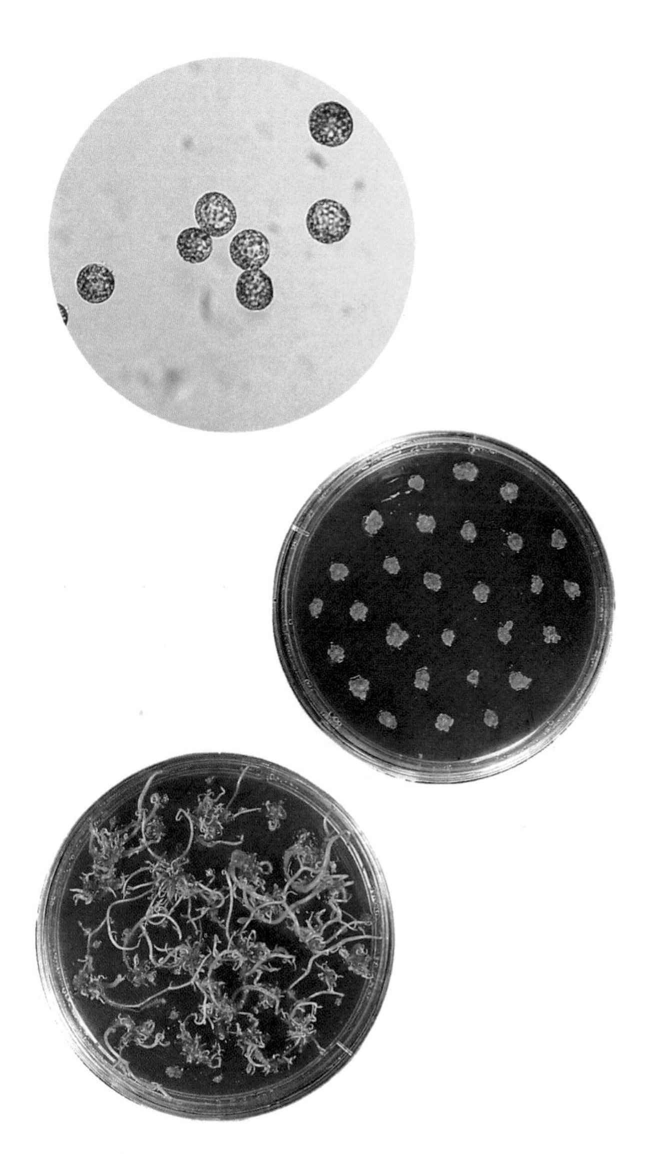

agriculture. Rapid advances in the understanding of plant cell biology have made it possible to grow a complete plant from only a single cell. In the past, if a new, desirable mutation occurred by chance, it was often years before enough plants could be bred to serve any useful purpose. However, one plant contains tens of billions of cells. If they can be separated and grown, the supply would be limitless. Even a single leaf could have enough progeny to supply the world for years. Many difficult and expensive cross-breeding techniques would be unnecessary.

By this method a single successful cross could be broken down, propagated and maintained forever. Equally, this method could short-circuit the long time it takes to build up desirable traits. If enough identical copies with one trait are manipulated, a second trait can rapidly be established and propagated. One dream is that soon there will be a potato-tomato hybrid which also fertilizes the soil

Cloning experiments with potato plants have involved taking cells called protoplasts from the leaves of one plant and placing them on a culture plate (top left). *Following many divisions, each protoplast produces a clump of cells called a callus* (middle left), *which eventually grows numerous shoots* (bottom left). *These plantlets are then potted and later planted out to produce a field of cloned potato plants* (above).

Fermentation units are used commercially to grow cultures of microorganisms for various biochemical products. Similar techniques can be used to manufacture gamma interferon (bottom).

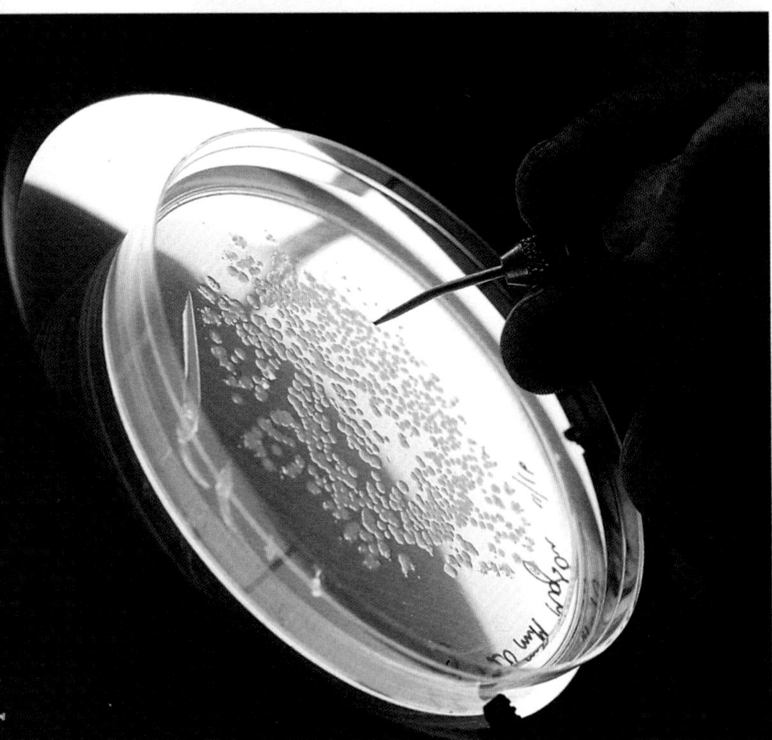

by fixing atmospheric nitrogen — a very efficient use of agricultural land.

Advances in test-tube fertilization and embryo implantation techniques have brought closer the previously only fictional possibility of cloned humans. Technically, all that is required is a woman prepared to donate egg cells, a sophisticated micromanipulator to perform the nuclear exchange, patience, and a lot of luck.

Recombinant DNA Becomes Big Business

Neither the scientists who developed the new biotechnology nor the world's capital markets have been slow to recognize the money-spinning potential of recombinant DNA. Because the whole library of heredity is now open, natural products — only just detectable a few years ago — can be readily produced by genetic cloning. A very wide range of substances can be made this way. Some, such as the recently-produced "tumor necrosis factor" (a potential step toward an eventual cure for cancer, and previously only suspected to exist), were earlier not really suitable because only tiny

The vast array of samples used for testing in the laboratories of Genentech, the giant American company at the fore of biotechnology, represents the many commercial uses of recombinant DNA.

amounts of impure material could be produced before the introduction of new technology.

The interferon story is a good example. When a cell has been infected with one virus it seems to produce a substance which prevents other adjacent cells from being infected by the virus. Something seems to be produced by the cell that interferes with the viral infection — hence the name interferon. For many years following the discovery of interferon there was fierce controversy about whether there really was any such substance. Eventually, at enormous cost, a very small amount was produced and although it was not pure, it at least convinced scientists that interferon was real.

Just as the available technology of the mid-1970s was being geared up at great expense to produce enough interferon from normal tissue for clinical use, genetic engineering appeared. Encouraged by people who believed interferon was the answer to everything from the common cold to cancer, several groups successfully cloned the interferon genes and used them to make previously unimaginable quantities. So many types of interferon are now available that its uses can be fully investigated. It shows promise, but it is not the miracle drug. The price of a single grain of interferon, estimated at 100,000,000 dollars in the mid-1970s (and never manufactured for that reason) has now dropped to about 10,000 dollars.

Drug firms rapidly realized the potential profits. Many multinational chemical firms recognized that genetic manipulation could solve many problems. And the scientists involved became aware that their advice and their graduate students were in great demand by industry. The result was a profusion of new genetic engineering and biotechnology companies. Many have now been bought or taken over by large firms anxious to get a foothold on the future; others have gone bankrupt. However, a few continue producing the genes and the gene products that are the basis of a whole new pharmaceutical industry.

Genetic engineers can make rare substances plentiful and relatively cheap. The prospects for other interesting proteins, especially scarce hormones and the lymphokines used by the body's natural defense systems, seem bright. A whole new industry has sprung up.

Chapter 6

Screening the Genes

Drawing water from the well was once a daily task for villagers. In a time of ignorance about genetics, deformed babies were thought to be caused by "bad air" or something in the local water supply. Modern genetic counseling acknowledges that there is an environmental factor in some congenital disorders, but most are caused by an anomaly inherited from one or both parents.

Genetic counseling involves the instruction of patients — particularly of prospective parents — in genetic principles and risk figures. In essence, it consists of giving as accurate information as is possible about the risks of transmission of inherited conditions, the means available to reduce such risks, and the treatment that is necessary or available to help anyone born with a serious inherited abnormality. Advice is sought not only regarding the risk of recurrence (although this is often the case), but also for genetic diagnosis, prognosis, or as a guide to treatment. As a recent World Health Organization report stated: "Genetic counseling is the most immediate and practical service that genetics can render in medicine and surgery."

People seek advice about various disorders, not all of which are inherited. A number have a reputation—warranted or not—for recurring in a kinship or a family, and for being caused by an alteration of the genetic material (chromosomes or genes). Most inquiries to genetic counselors come from parents seeking advice on the risk of recurrence for themselves, their children, or their grandchildren, representing an anxiety that between family members often remains unspoken, but an anxiety that is nevertheless real. Other inquiries come from relatives of patients with genetic disorders, or persons who are concerned about an apparent high frequency of common conditions such as coronary heart disease or cancer among their relatives. Many individuals who seek genetic counseling have had the worst picture painted for them and may have groundless fears, but they all have in common the need for accurate, detailed information and help.

It has been estimated that about four per cent of the world's population suffers from some genetic, or partly genetic, abnormality, and would therefore benefit from genetic counseling. At least one per cent of all infants, for example, have a major chromosome anomaly. In some regions of the

Various blood disorders, such as thalassemia, are genetic in origin. In the photomicrograph of normal blood (left), *red cells predominate, whereas they are absent from the sample of anemic blood* (right).

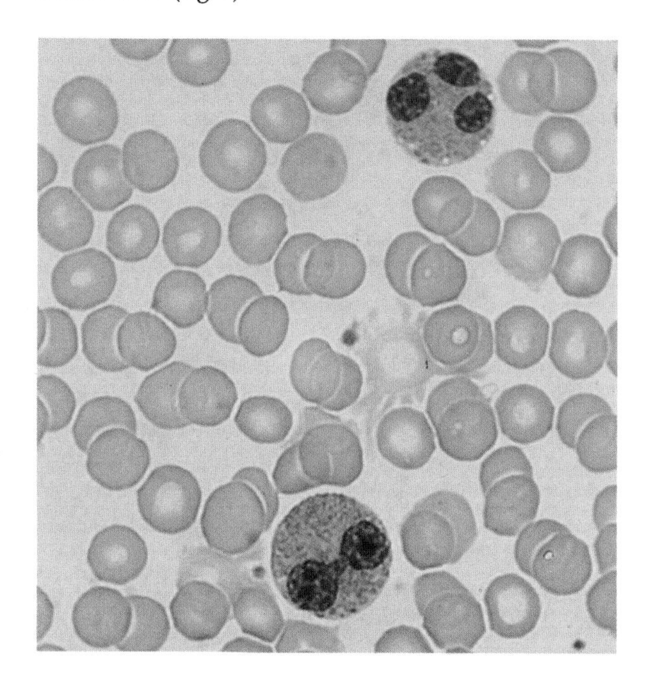

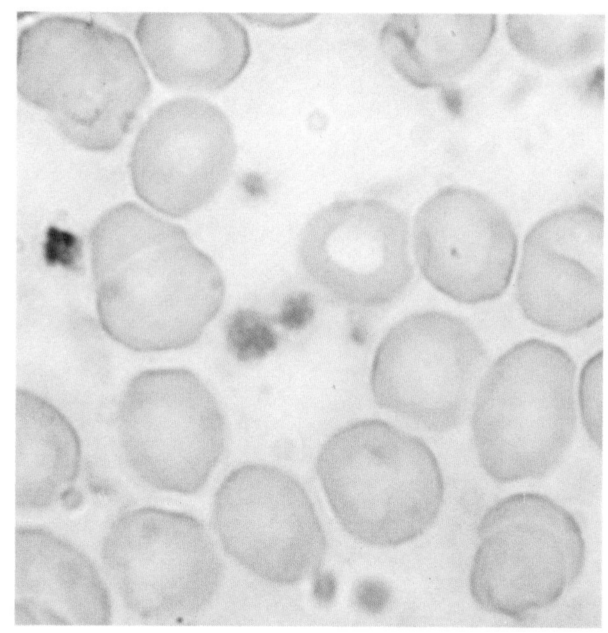

world, the proportion of persons suffering from a genetic disorder is much higher. In tropical Africa, for instance, there are certain areas where the gene for sickle-cell anemia is carried by one-fourth or more of the population but where genetic advice or counseling is rarely available. In Italy, conversely, for the detection of thalassemia there is a network which is based on an Institute in Rome, with provincial centers that provide the local people with facilities for accurate and detailed diagnosis and premarital counseling.

Recent advances in medicine and surgery, along with increasing public health measures such as the purification of water and the supervisory control of diet, have gradually led to the control of disorders and disabilities caused by nutritional problems and infectious diseases. Gradually, with time, environmental disorders are being replaced by those which are largely or partly caused by genetic factors. It is estimated that approximately one in twenty children admitted to the hospital has a disorder which is entirely genetic in origin, and that such disorders account for about one in ten of childhood deaths in the hospital.

The true frequency of genetic disease is difficult to estimate precisely. Many conditions that are clearly and strongly genetically determined and inheritable are rare. But there are enough different kinds to be relatively common in aggregate. Furthermore, some of the relatively common single-gene defects are distributed unevenly among different ethnic groups. Single-gene disease, caused by mutant genes of large effect, may occur in as many as eight per thousand births, and malformations in newborn babies caused by single genes in about half this figure.

Gathering Statistics

The frequency of chromosome disorders at conception can be estimated most accurately by considering karyotype analyses from early human embryo data (the karyotype is the chromosomal constitution of the cell nucleus). Most estimates are based on such investigations of spontaneous abortion, stillbirths and newborn babies. In the past, X chromosome surveys have had the advantage of simplicity, but provide information only about the numerical variation of the X chromosome and not the other chromosomes.

The incidence of chromosomal aberrations or abnormalities in karyotypes in consecutive newborn surveys from a number of countries is close to about one per cent, if all major, as well as minor, abnormalities are included. Numerically, the most

Potato blight, cause of famines in nineteenth-century Ireland, was once also blamed for disorders such as neural tube defect. It is now known that the cause of the defect can be a genetic abnormality.

common abnormalities are the syndromes associated with the autosomes, or non-sex chromosomes. These include trisomy 21 (Down's syndrome), trisomy 18 (Edwards' syndrome) and trisomy 13 (Patau's syndrome). The syndromes commonly associated with anomalous X or Y chromosomes are Klinefelter's syndrome (which is XXY genotype) and Turner's syndrome (which is XO). In addition there are structural abnormalities involving the autosomes, including the so-called cri-du-chat, or cat-cry syndrome, and absence of chromosome number 18.

A high incidence of chromosomal abnormalities is seen in several disease categories. The most significant are early spontaneous abortions (sixty per cent), major multiple congenital malformations (five to twenty per cent), infertility and sterility in various groups of patients (one to ten per cent), and mental retardation (one to three per cent). In most pregnancies, the chromosomal abnormalities occur as chance random events. These situations are very unlikely to recur, because the parents' chromosomes are normal.

There is, however, a small but important minority of cases in which one parent is the carrier of a chromosome rearrangement (translocation) but is otherwise normal, or in which one parent is a mosaic: in his or her tissues there is a small proportion of cells with a different chromosome number. In such families, the recurrence risk may be up to ten per cent or higher.

It is well known that there are marked ethnic and geographic variations in the frequency of genetic disease. In the United Kingdom, for example, the frequency of neural tube malformations varies markedly from 1 to 1.5 per thousand total births in south-eastern England, to 1.5 to 2 per thousand in the Midlands, increasing to 2 to 2.5 per thousand in northern England, 2.5 to 3 per thousand in Scotland and Wales, and 4 per thousand in Northern Ireland. If someone from Wales, for example, moves to Los Angeles, he or she tends initially to take along the Welsh risk figure of having a child affected with neural tube defects. But after arriving in Los Angeles, the risk of having an affected child falls to about halfway between that of a person living in Los Angeles and that of someone in Wales. Conversely, if someone living in Los Angeles moves to Wales, the risk of having an affected child with neural tube defects increases from what it was previously in the United States, but certainly does not increase to that of a person who grew up and lives in Wales.

When both inheritance and environment — geography in this case — are involved in causation, the term multifactorial is used — that is, it involves many factors. Multifactorial origin is postulated for many major malformations. It was once suggested, for instance, that there was a relationship between potato blight and neural tube defects. For a counselor, sometimes to disentangle familial clustering caused by genes in common from that resulting from a shared environment, not to mention any interaction, can be extremely difficult. Furthermore, the frequency of some multifactorial conditions changes rapidly, whereas the frequency of others appears to be fairly constant. In ischemic heart disease, the frequency of death has increased markedly since the 1920s in Europe. And certain African populations have rapidly acquired European frequencies for ischemic heart disease once they have adopted a European diet and way of life. No consistant changes in the frequency of incidence have occurred with the more common congenital malformations.

Reasons for Counseling

There are two principal reasons for patients to seek genetic counseling. The first is to obtain advice following the birth of an affected child, and the second is to determine their risk as part of a specific population with relation to a particular condition. Down's syndrome is a chromosome disorder manifested as an increased number of chromosomes (47 instead of 46), termed trisomy 21. It illustrates both reasons why parents seek genetic advice. The British researcher Lionel Penrose showed that there was a relation between increasing maternal age and Down's syndrome. At one time, there were numerous theories attempting to account for this correlation, including the idea of a worn-out uterus and the fact that there was also a relation between an increasing number of pregnancies and the likelihood of having a child with Down's syndrome.

Down's syndrome is the most common single

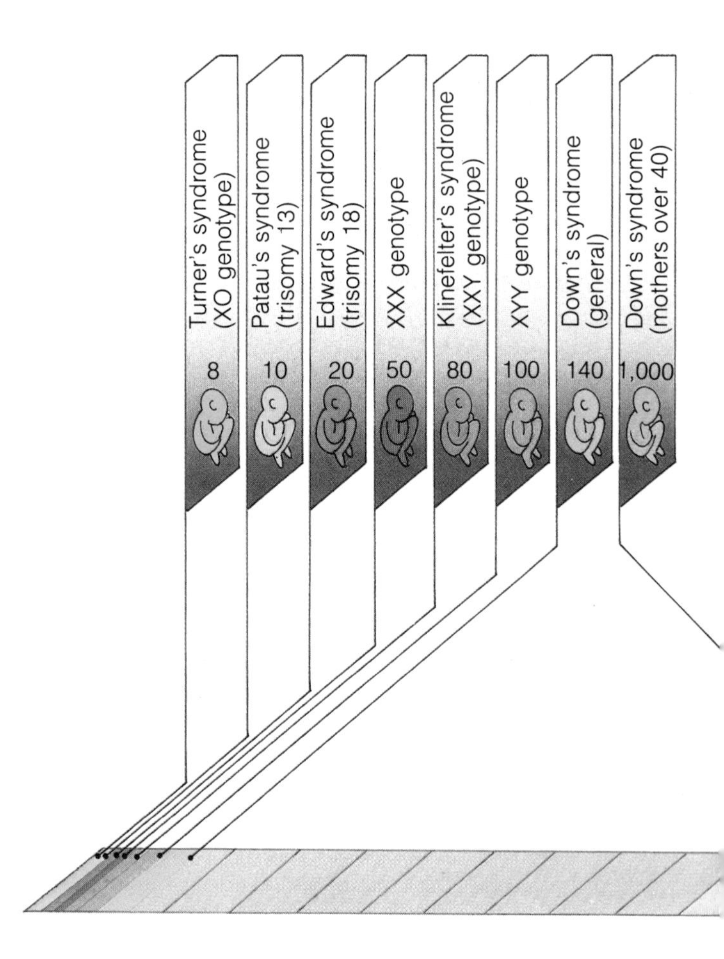

cause of mental retardation, acccounting for an estimated thirty per cent of all children with severe mental handicap. Recent studies have suggested that the prevalence of the disorder is increasing. The risk of giving birth to a child with Down's syndrome (as well as other chromosome abnormalities) increases rapidly in women more than thirty-five years old. At age thirty-five the risk is about 1 in 300, at age forty it is 1 in 100, and at forty-five it is about 1 in 50. But chromosomes studies after amniocentesis (involving an analysis of amniotic fluid) reveal that the risk of having a child with trisomy (of which two-thirds are Down's syndrome) is about 1.5 to 2 in 100 for mothers between the ages of thirty-five and thirty-nine, and for mothers of forty years of age or older it is 5 in 100. Amniocentesis data is of more recent origin than live-born data, and for this reason prenatal figures for chromosome abnormalities may reflect postnatal figures more closely. Also, there may be a genuine increase in the age-specific incidence of Down's syndrome, with the effect being most marked in women from the thirty-five to thirty-nine age group.

In Britain and West Germany, about 1.5 per cent of all births occur to mothers aged forty or over.

This percentage is falling, but these births account for nearly seventeen per cent of all patients with Down's syndrome. Mothers of thirty-five years of age or over account for eight per cent of all births, and about thirty-seven to forty per cent of all children with Down's syndrome. Therefore, without even knowing the cause of Down's syndrome, a counselor can identify a small population at risk — namely, women aged thirty-five years or more who are pregnant — and offer them a prenatal diagnosis by means of amniocentesis or chorionic villus biopsy in some centers. Theoretically, in those countries where abortion is allowed, the birth of about thirty-seven to forty per cent of all children with Down's syndrome could be prevented. This analysis leaves roughly another forty to forty-five per cent of patients with Down's syndrome born to mothers who are less than thirty-five years old. At present there is no way to identify the women who obviously are at risk, and to whom amniocentesis could be offered — unless they are proved to be carriers of a translocation, or if for some unrelated reason they are discovered to have decreasing levels of blood alphafetoprotein.

Not all disorders that affect more than one member of a family are genetic in origin. In fact, a

Analysis of birth statistics by genetic counselors reveals the incidence of various genetic disorders. The folded band in the diagram (above) *represents 100,000 births, of which the great majority (shown in brown) are normal. About 1,000 babies out of every 100,000, born to mothers over forty years of age, have Down's syndrome (shown in green). Another 140 per 100,000 babies, born to younger women, also have Down's syndrome. Other chromosomal aberrations are more rare, ranging from 100 per 100,000 for the XYY genotype to only 8 per 100,000 for the XO genotype, also known as Turner's syndrome.*

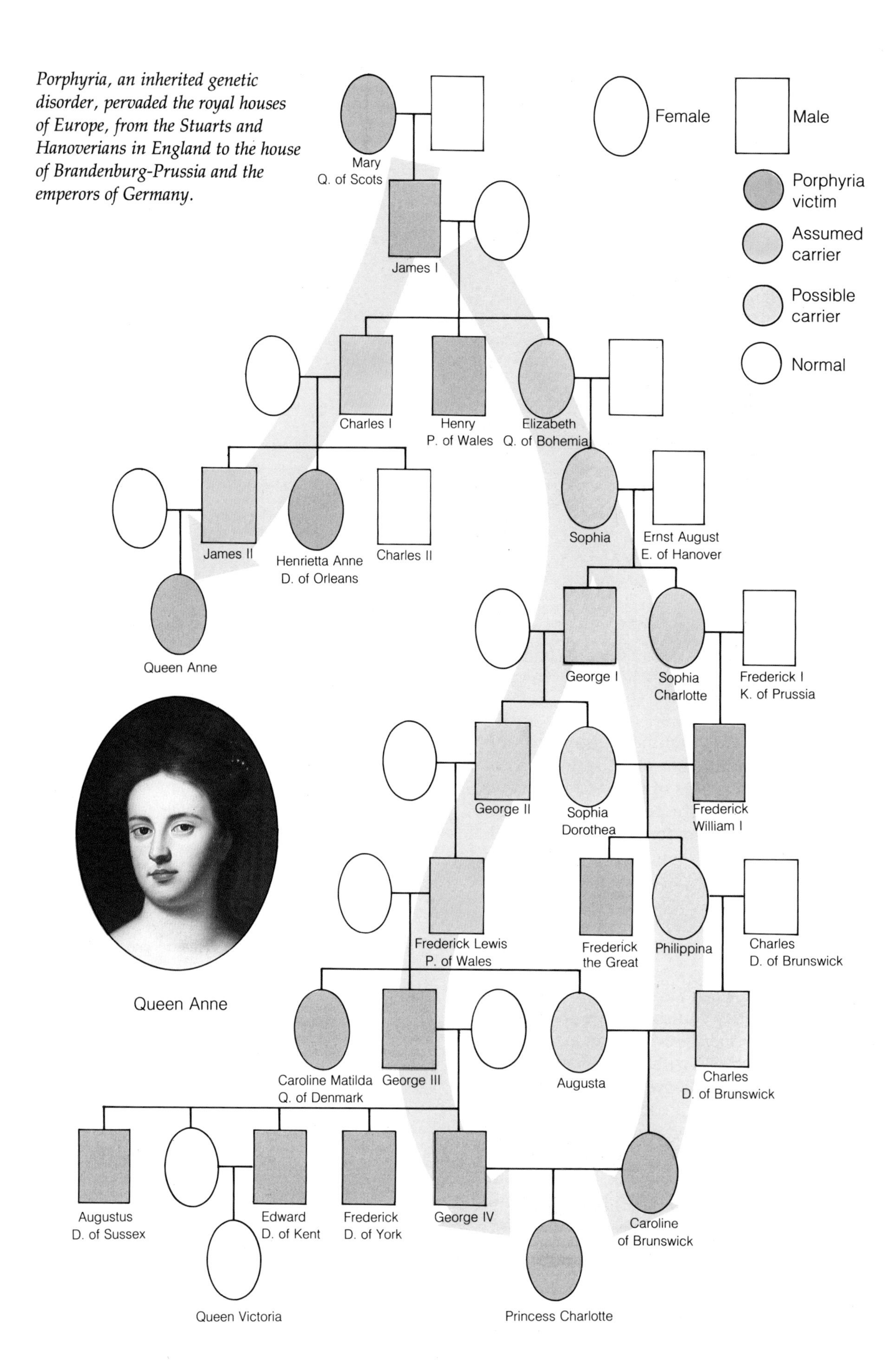

Porphyria, an inherited genetic disorder, pervaded the royal houses of Europe, from the Stuarts and Hanoverians in England to the house of Brandenburg-Prussia and the emperors of Germany.

definable environmental cause — for example, a drug, X rays or a virus infection — may affect more than one member of a family at a time. For this reason, a key factor in genetic counseling involves investigating the family's medical history. A genetic history is a compilation of all information relevant to a patient's condition. Many hospital family histories, for example, are inadequate because they record information only with respect to a few disorders such as heart disease, cancer, diabetes or "mental retardation," which may have little or no connection with the patient's condition. The exact relationship of the relatives to the propositus (or index case) and to one another has to be established, and enable the counselor to construct a pedigree chart or family tree. The pattern that a genetic disorder exhibits within a family may suggest a diagnosis and indicate diagnostic tests to confirm it.

Hemophilia is a disorder in which there is an inborn tendency to bleed because of a defect in the clotting ability of the blood. The pattern of inheritance, which it shares with other disorders resulting from genes on the X chromosome, was recognized more than a thousand years ago. According to the Talmud, boys whose brothers were bleeders were exempt from circumcision, as also were the sons of sisters of women who had male bleeder children. The exemption did not extend to fathers' sons by other women, making it clear that the inheritance pattern of the disease was already understood at that time. A number of the descendants of the British Queen Victoria are known to have died from hemophilia, and it is interesting to speculate on a "what if" basis about what, if carrier detection and prenatal diagnosis had been available to Queen Victoria, might have happened to the monarchy.

Another disorder which illustrates the importance of preparing a careful family tree is porphyria, which has in the last centuries affected the royal families of Britain and Prussia. Porphyrias are hereditary errors of metabolism. The porphyrins are purple-red pigments which are contained within every cell of the human body. They are responsible for the red color of blood by their presence in hemoglobin in red blood cells.

Normally the body retains most of the por-

King George III of England, known to his critics as "Mad King George," suffered from a severe form of porphyria, which he inherited from his ancestors as shown on the "pedigree," or family tree, illustrated opposite. At times his mental disturbances were so severe that he had to be restrained in a straitjacket; meanwhile the nation was effectively without its leader.

phyrins, and only small amounts are excreted in urine and feces. In the porphyrias, large amounts of porphyrins are formed and excreted, giving the urine a characteristic port-red color. Their accumulation in the body produces other symptoms including abdominal pain, vomiting and constipation, paralysis and mental manifestations, in addition to skin sensitivity to light. The porphyrias are inherited as Mendelian dominant traits; more than one form exist.

Recently, two direct descendants of the English King George III have been tested and shown to have porphyria. Once such a condition has been demonstrated in even only one living member of a family, a historical analysis tends to fall in line with genetic studies, which trace the condition from an established living individual back to former generations. This is the equivalent of performing the test on George III himself, since a coincidental occurrence of such a rare genetic disorder can be discounted. Recent studies have derived clinical data spanning thirteen generations over more than four centuries, documenting this rare inborn error of metabolism.

Throughout history, many important historical figures have been diagnosed as having kidney problems or gout. The physicians in George III's time "had no stethoscope, no knee-jerk hammer, not even a clinical thermometer." In addition, some commentators have stated that the physicians never examined him and could not even speak to the patient unless the patient first spoke to them. The historical significance of the diagnosis is that it caused directly or indirectly three major national disasters for Great Britain. The first, caused largely by George III's ineffective and spasmodic leadership, came to a climax in 1775 when the thirteen American colonies rebelled against British rule; the final result was the founding of the United States. The other disasters were the Regency crisis in 1788, when George III had his most severe attack, and the catastrophe of 1817, when Princess Charlotte died in childbirth with her infant. Her death threatened the Hanoverian succession with extinction and left Britain without an heir apparent until the birth of Queen Victoria (in 1819). Even earlier, porphyria may have contributed to the fact that after seventeen confinements Queen Anne left no heir. This necessitated safeguarding the succession by transferring the Crown from the House of Stuart to that of Hanover. As a result, George I and his descendants — carriers of porphyria — succeeded to the English throne.

Most of the Royal Houses of Europe of the eighteenth and nineteenth centuries can trace their descent through James I. They include the House of Hanover; the exiled family of the Stuarts; the House of Brandenburg-Prussia, who in the nineteenth

The funeral of Princess Charlotte (left) *took place at night. Daughter of King George IV, who married his cousin from Brunswick, she inherited porphyria and died in childbirth at the age of twenty-one.*

Gamblers in a Monte Carlo casino try their luck at games of chance, attempting to beat the odds and win against the "bank." Much of genetic counseling involves calculating odds and probabilities, estimating the risks that a genetic disorder might occur in a pregnancy. Often predictions are based on a family history of the prospective parents. During pregnancy, prenatal diagnosis can give exact information.

century became Emperors of Germany; the House of Orleans; the House of Savoy, who became Kings of Italy; and the House of Orange, who became Kings of Holland. Porphyria has been diagnosed in a Queen of Scotland of the sixteenth century, and a King of Great Britain of the nineteenth century. There is no evidence that Queen Victoria suffered from or transmitted porphyria to her descendants, but she was almost certainly a carrier for hemophilia. Hemophilia and porphyria together are the royal maladies which have affected the Royal Houses of Europe, and in turn the history of Europe and, indirectly, the United States.

The Basis of Genetic Counseling

Genetic counseling and advice has to deal mainly in statistical probabilities. The giving of genetic advice is fairly straightforward in typical conditions. This is particularly true for the three major types of simple Mendelian inheritance, namely autosomal dominant (without complications introduced by considering variations in effect), autosomal recessive (in which risks to carrier siblings are small unless they marry a near relative), and X-linked (generally recessive with risks to the male children of carrier women).

When the specific diagnosis and pattern of transmission of a trait is known, the risk of recurrence is usually simple to calculate. For example, cystic fibrosis, in which about one in every twenty or twenty-five individuals is a carrier of the altered gene, is an autosomal recessive with a one in four risk for subsequent children if both parents are carriers. Classic hemophilia is an X-linked recessive in which the risk that any son of a carrier woman will be affected is one in two (an even chance), and the risk that a daughter will be a carrier is also one in two.

When a trait is inherited multifactorially, genetic counseling can usually be provided without too

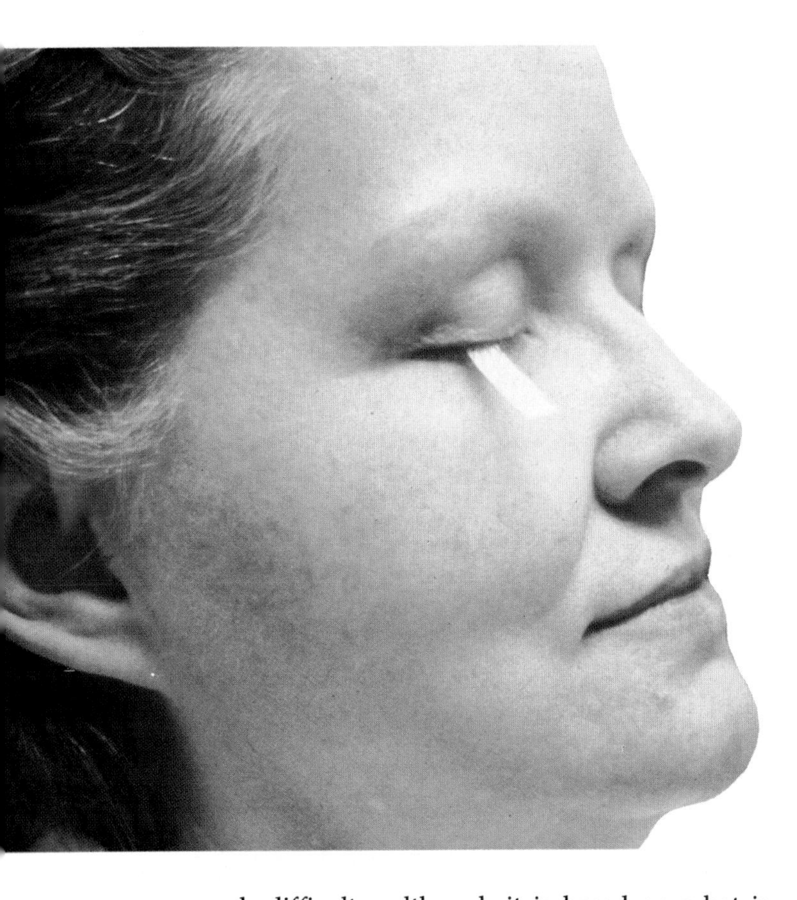

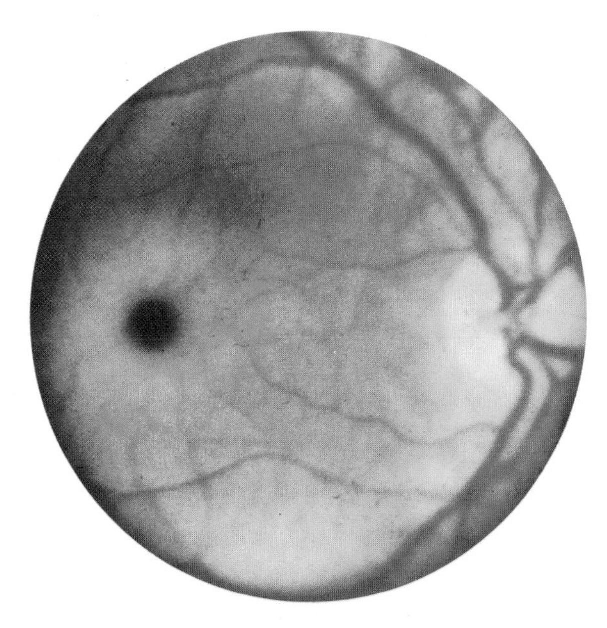

much difficulty, although it is based on what is called the empiric risk that it will recur — that is, on the basis of past experience of the behavior of the trait within families. Although empiric risk figures are quite useful guides, they are only general averages based on experience and do not necessarily constitute an accurate estimate of the risk for any individual in a specific family. At times the counseling can be made difficult simply because in some disorders risks are extremely difficult to estimate. Fortunately, the more uncertain the hereditability of the condition, the lower is the risk of its recurrence.

Three specific points emerge from these facts. First, the formulation of genetic outlook, or advice, demands a precise diagnosis with an adequate family history that takes into account all known dominant traits. Secondly, a counselor has always to remember the heterogeneity of genetic disorders. Heterogeneity implies that the manifestations of the disorder can be caused by different genes. This situation is sometimes revealed by laboratory investigations or sometimes by the family history. Thirdly, there is the question of how to assess what constitutes a high or grave risk and, conversely, a low or slight risk.

In practice genetic counselors distinguish "low risks" from "high risks," but the distinction is entirely pragmatic. This generalization regarding risks is based on the counselor's scientific assessment of the situation, rather than the fears or anxieties of the persons seeking advice. Assessment of the degrees of risk is usually based on the fact that any woman in any pregnancy has a one in forty to one in thirty risk of having a seriously or moderately malformed child. This is the risk all potential parents face whenever they want children. Against this background figure, therefore, the counselor places a risk of one in ten or higher and defines it as a high risk, or a risk of one in twenty or lower, which is regarded as average or low. Fortunately, very few risk figures fall between these values. In counseling it has been shown that on average about two-thirds of the estimates of risk given are in the low-risk figure.

As for the accuracy of genetic counseling, few statistics are available. One recent study, however, showed that out of ninety-four children born after consultation to "high-risk" couples who decided to go ahead and have children, seventeen (or about 18.1 per cent, one in five) were abnormal, whereas 229 "low-risk" couples produced only five specifically affected children (2.2 per cent or about one in fifty). Other studies confirm the findings, and it would seem that genetic advice is worth taking.

The discussion so far has dealt only with risks

Tay-Sachs disease involves an enzyme deficiency, which can be detected by sampling tears on absorbent paper (left). *Its symptoms include dark spots on the retina leading to blindness.*

evaluated in terms of probabilities. To make the statistical meaning clear, there are several analogies from gambling such as tossing a coin (to demonstrate the one to one risk of autosomal dominant disease to siblings or of transmission by affected parents), or the throwing of a dice (to explain autosomal recessive risks to siblings), or the drawing or a given card from a deck (to explain empirical, multifactorial risks).

Ultimately, couples contemplating having another child after the birth of an affected one, pregnant women, or women about to be married want direct rather than statistical reassurance. Preferably they would like reassurance about any abnormality, but certainly need it about any specific high risks or even moderate risks which they themselves face. Whether a woman considers pregnancy or not, or is prepared to continue a pregnancy at the time of seeking advice, depends on the nature and size of the risk. Generally in only about one in twenty low-risk pregnancies do women insist on termination, although they do so in well over one-third of high-risk situations. Direct as opposed to probabilistic counseling may change the picture completely — and this entails obtaining accurate information through prenatal diagnosis.

Prenatal Diagnostic Techniques

Investigations during pregnancy follow a number of different approaches, all of which are directed at diagnosing specific fetal abnormalities. When an abnormality is present it may be treatable and perhaps cured; alternatively, the pregnancy may be terminated if the condition is grave and untreatable. Advice has to be in keeping with the anxiety that such knowledge generates in the parents and conform with their wishes at a time in pregnancy when termination — if morally and legally acceptable — is still practicable. A useful distinction is drawn between investigations made in later pregnancy and those done early, say within the first three to four months.

The recent introduction of specialized techniques for prenatal diagnosis of various genetic defects represents a new approach in medicine and has added a new dimension to genetic counseling. It is now possible, in counseling for some conditions, to say definitely that a particular fetus is, or is not,

A woman in a Moscow street ponders whether or not to buy lottery tickets from a street vendor. It is easier to calculate the chance of winning a prize in a lottery than it is to estimate the probability that someone will be born with a multifactorial genetic disorder, in which several genes and environmental factors act together.

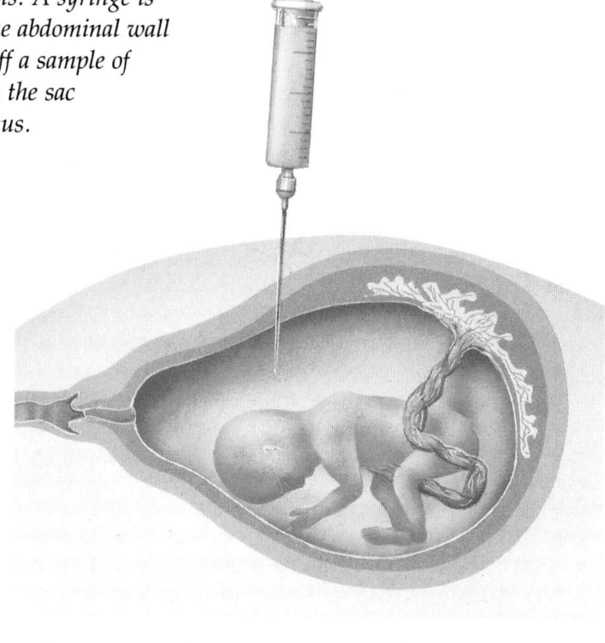

Amniocentesis is a major technique in prenatal diagnosis. A syringe is inserted through the abdominal wall and used to draw off a sample of amniotic fluid from the sac surrounding the fetus.

affected, and to do so at a stage early enough to allow selective abortion (where permissible) of fetuses that are abnormal.

Prenatal diagnosis, then, although also aimed at diagnosing disorders for which prenatal treatment is available, focuses predominantly on detecting "seriously handicapping" physical or mental abnormalities which, because they are untreatable or incurable, may justify termination of pregnancy. It is therefore not surprising that prenatal diagnosis, screening and the notion of selective abortion should generate discussion from all areas. The ability to carry out prenatal diagnosis and screening is an outcome of various developments in medical genetics. These include the growing of human cells in small plastic bottles in tissue culture, analyzing the chromosomal makeup of such cells, analyzing the activities of many of their enzymes, immunological testing of the amniotic fluid itself, and even analyzing specific genes.

The developing fetus is enclosed in a sac made of two membranes, the inner amnion and the outer chorion. The sac itself is filled with amniotic fluid, which is formed by the fetus and consists mainly of urine and secretions from the respiratory tract. The amniotic fluid also contains the cells which are sloughed off from the fetal skin and other cells from the lungs and bladder. They are collectively known as amniotic fluid cells. The best established techniques involve removal of a sample of amniotic fluid through the abdominal wall at about sixteen to seventeen weeks after conception, followed by culture of the amniotic fluid cells, or various chemical tests on the fluid. It should be emphasized that there have to be strict medical reasons for carrying out the procedure. Although another condition, which was not anticipated, is sometimes discovered, the object of the test is to ascertain whether or not the fetus is normal.

In screening, amniocentesis is used in two ways. First it is employed in the narrow sense of testing a fetus whose probability of having a genetic disorder is known, from a knowledge of the genotype of the parents. Secondly, amniocentesis is used in the general sense of screening for reproductive information, and if an anomaly is detected, offering a defined population the choice of allowing an affected fetus to survive.

Down's syndrome is a relatively common subject for both these uses of prenatal diagnosis through screening by amniocentesis. Thirty-five to forty per cent of all Down's syndrome babies are born to mothers over thirty-five years of age. Thus if such women were to request prenatal screening by amniocentesis followed by termination of abnormal fetuses, the incidence of Down's syndrome would be reduced by approximately thirty-five to forty per cent. Yet a significant number would remain undetected by current screening practices, namely those Down's syndrome babies born to mothers under thirty. Why shouldn't screening by amniocentesis be extended to all pregnant women, then? There are practical problems, such as the lack of facilities and personnel. In addition, it is still not known whether the risks of the procedure — including that of a miscarriage — outweigh the benefits. And there is also the question of ethics.

Additional methods of prenatal diagnosis which have now become available include chorionic villus biopsy — examination of a tiny section of the chorion — and the use of a fiberoptic device, or fetoscope. This instrument allows direct visualization of the fetus and provides a means of obtaining small samples of pure blood from the fetus for the prenatal diagnosis of blood disorders such as hemophilia. A fetoscope can also be used for carrying out direct intrauterine surgery.

Problems Associated with Screening

The first problem is the abortion decision. Most workers take the view that amniocentesis and

The fluid sample from amniocentesis is processed and analyzed to provide information about the fetus. A centrifuge is used to separate fetal cells from the fluid part, and each is subjected to various tests. Some cells may also be cultured to provide information about the sex of the fetus and its chromosomal makeup.

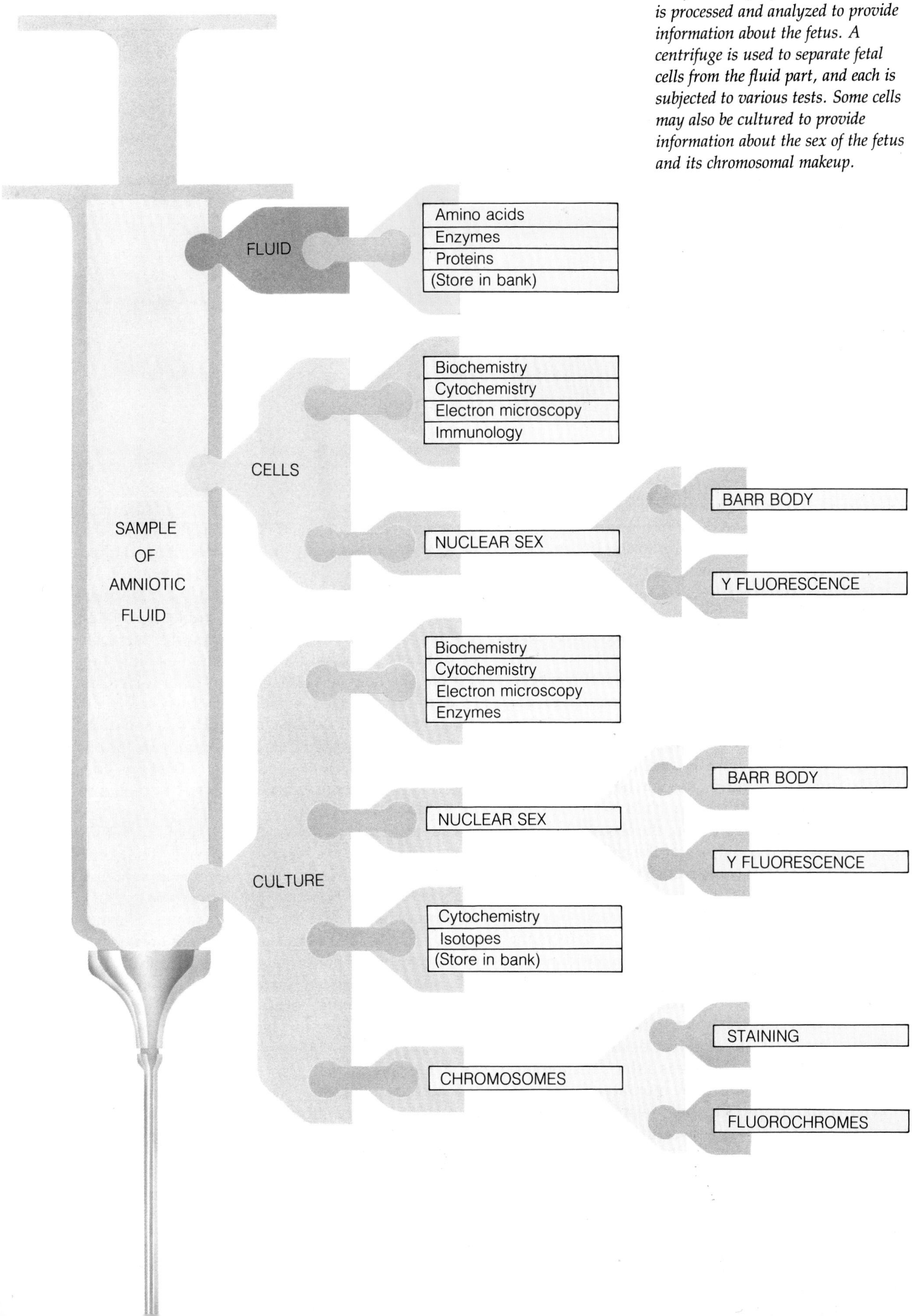

SAMPLE OF AMNIOTIC FLUID

- FLUID
  - Amino acids
  - Enzymes
  - Proteins
  - (Store in bank)
- CELLS
  - Biochemistry
  - Cytochemistry
  - Electron microscopy
  - Immunology
  - NUCLEAR SEX
    - BARR BODY
    - Y FLUORESCENCE
- CULTURE
  - Biochemistry
  - Cytochemistry
  - Electron microscopy
  - Enzymes
  - NUCLEAR SEX
    - BARR BODY
    - Y FLUORESCENCE
  - Cytochemistry
  - Isotopes
  - (Store in bank)
  - CHROMOSOMES
    - STAINING
    - FLUOROCHROMES

Hemophilia is an inherited disorder in which the blood fails to clot properly after injury. A baby's blood is tested by measuring the time it takes for a sample to clump and form clots (below right).

Chromosomes can be seen dividing during meiosis (right), *the special type of cell division that gives rise to sex cells (eggs and sperms). Incorrect splitting can result in genetic anomalies in a fetus.*

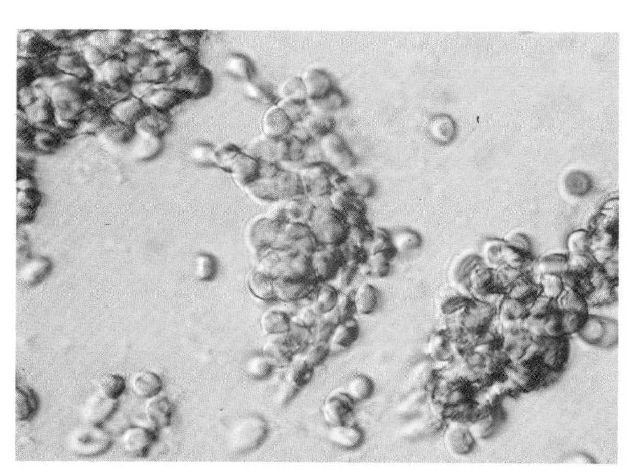

abortion must be considered as two separate issues — that is, a woman should not be committed to agree to an abortion (where permissible) before the anmiocentesis is carried out. For how are parents to know in advance their reaction to the information that the fetus is abnormal? In these days of diminishing numbers of babies available for adoption, there are now cases of pregnancies brought to term despite this knowledge. The social and psychological impact of this knowledge on its recepients requires study. Too little effort has been spent so far in the detailed examination of the thoughts, feelings and attitudes of women who have undergone amniocentesis, or indeed of their partners. In a recent famous case, a woman whose obstetrician did not inform her that her fetus had an abnormality gave birth to a child with Down's syndrome. After a court action, the obstetrician was made to pay for the support of the child.

It is clear that if one takes the absolutist position that under no circumstances is abortion morally defensible, then prenatal diagnosis has no place in medical practice. However, if there is agreement that the fetus is not the only one to be considered in making such a decision, and the welfare of the mother, the family, or society at large are also to be taken into account, then abortion can become an ethically accepted method of dealing with an abnormal fetus. Clearly, there is a sharp dichotomy of view on the question of the morality of abortion in general and its implications toward pregnancies in which there is a clear risk of an abnormal fetus. Each side must ultimately go its own way.

A second problem is the promotion of screening. When, if ever, should prenatal screening for Down's syndrome be publicly promoted like other screening tests (such as cervical smear tests)? There is a vast difference between recommendations made between the physician and the patient in the privacy of the consulting room in the traditional patient–doctor relationship, and public promotion via the government, the media, or social ostracism of affected persons and their families. In other words, the question is: should prenatal diagnosis by amniocentesis for women over thirty-five years of age be pressed publicly by effort, or should the educational effects be limited to notifying the public of its availability and reminding physicians that they have a responsibility to make it available to their patients?

The third problem concerns unexpected findings. Here the problem is twofold. First, the unexpected finding of other chromosomal abnormalities, such as trisomy (other than trisomy 21), abnormalities of the sex chromosomes, balance translocations, chromosome fragments and mosaicism. Such abnormalities can be found when the anmiocentesis is carried out to screen for Down's syndrome, and equally when the procedure is done for biochemical indications, screening for spina bifida, or even sexing the fetus for X-linked disease.

Most of them pose no problem because the cytological abnormality is associated with a lethal or severe disorder, in which the fetus dies before birth. However, the abnormalities of the sex

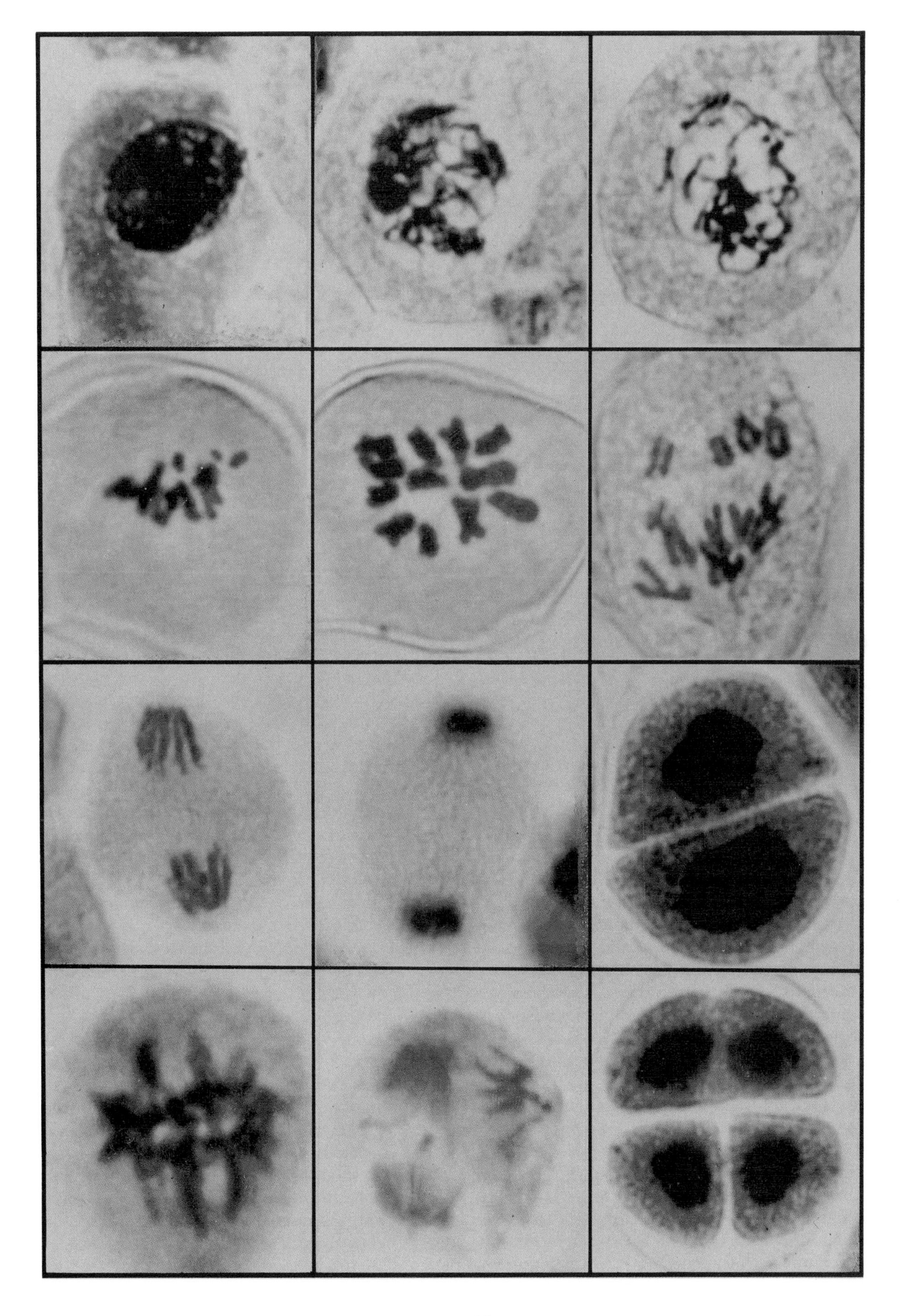

A radioimmuno-assay (RIA) test is carried out on a small sample of blood from a newborn baby (below) *to detect possible abnormalities, such as a defect in the metabolism of one of the essential enzymes.*

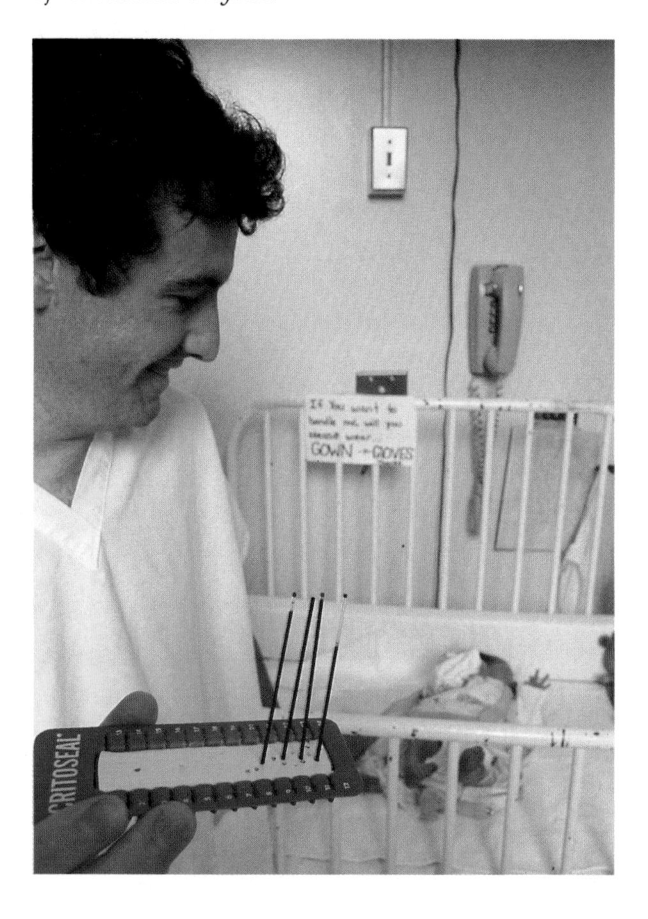

chromosomes do pose a problem. The counselor cannot state that the fetal karyotype is normal, so there appears to be no alternative but to tell the parents of the results. Because there is only the karyotype of the fetus to go on, there is a problem in conveying the clinical information about the variability of severity and expectations for persons with Klinefelter's syndrome, or the XYY syndrome, to parents to whom the event is entirely unexpected. They may know nothing at all of the condition or, worse, may believe that such individuals must end up at odds with society. To deliberately withold this information from parents would therefore be ethically unjustified.

The second problem concerns twins. The incidence of fraternal twins — like that of Down's syndrome — also increases with maternal age. The likelihood of the discovery that one twin is normal and the other twin has Down's syndrome, or another unexpected abnormality, will increase as more women are offered the opportunity of having prenatal diagnosis either by amniocentesis or by the more recent technique of chorionic villus sampling. This latter method has the advantage that it can be performed early in the pregnancy, during about the fourth week, whereas amniocentesis is usually reserved for high-risk mothers and is carried out at between fourteen and sixteen weeks of pregnancy. But the dilemma still exists: what if one twin is healthy and one is not?

A much more immediate problem arises from the enormous range of variation in the severity of the genetically determined conditions that can be identified by prenatal diagnosis, or which are likely to become candidates for prenatal diagnosis. The issues seem clear when the fetus is diagnosed as being affected with Tay-Sachs disease. Here a program for screening for Tay-Sachs disease, which has a low frequency in the general population but a relatively high frequency in certain Jewish groups, has been operating successfully in the United States and England.

Tay-Sachs disease, an autosomal recessive condition involving an enzyme defect, is especially prevalent among Ashkenazi Jews. It is invariably fatal by the age of four and there is no known cure. The affected gene is ten times more likely to be present in Ashkenazi Jews than in other people, and so the disease itself is about one hundred times more frequent in these Jewish populations than elsewhere. If two prospective parents are found to be carriers, then prenatal diagnosis can be carried out, fetal cells grown in culture, a specific test done for the enzyme defect, and if the fetus is affected a termination can (where permissible) be carried out. The risk in each pregnancy of a couple who are carriers for Tay-Sachs disease is one in four.

In the past, blood has been taken in mass sessions at such places as colleges or synagogues, for example. Five milliliters (about a teaspoon) of blood is drawn and tested for the presence of the enzyme hexosaminodase A. Carriers of the Tay-Sachs gene have an appreciably lower level of this enzyme in their blood serum than noncarriers. Recently, it has been shown that it is more effective and efficient to screen utilizing tears, in which the level of hexosaminodase A is about fifteen times as high as it is in the blood. By using tears, screening

Some babies are born with an enzyme defect that prevents them from digesting the sugar galactose, which is a normal constituent of mother's milk and cow's milk. The condition, galactosemia, is easily treated by feeding the baby a special galactose-free formula.

becomes much more acceptable and does away with the problems and difficulties of obtaining and storing blood.

The issues are not, however, as straightforward with genetically determined abnormalities which previously resulted in severe disability, but for which there are now effective forms of treatment. The situation has already arisen with galactosemia which, if left untreated, results in mental retardation, eye cataracts, liver damage, and usually death in childhood. Prenatal diagnosis can be achieved by a specific enzyme test on amniotic fluid cells grown in tissue culture. At the present time, pregnancies at risk can be identified only after the birth of a previously affected child. Effective treatment requires early diagnosis in the postnatal period, and a feeding schedule for the baby involving a galactose-free diet.

The early detection of some forms of spina bifida can now be carried out by testing a small sample of

Prenatal diagnosis is increasingly used as a part of genetic counseling. Particularly useful is a combination of amniocentesis and ultrasonic scanning, here being carried out on a woman in early pregnancy.

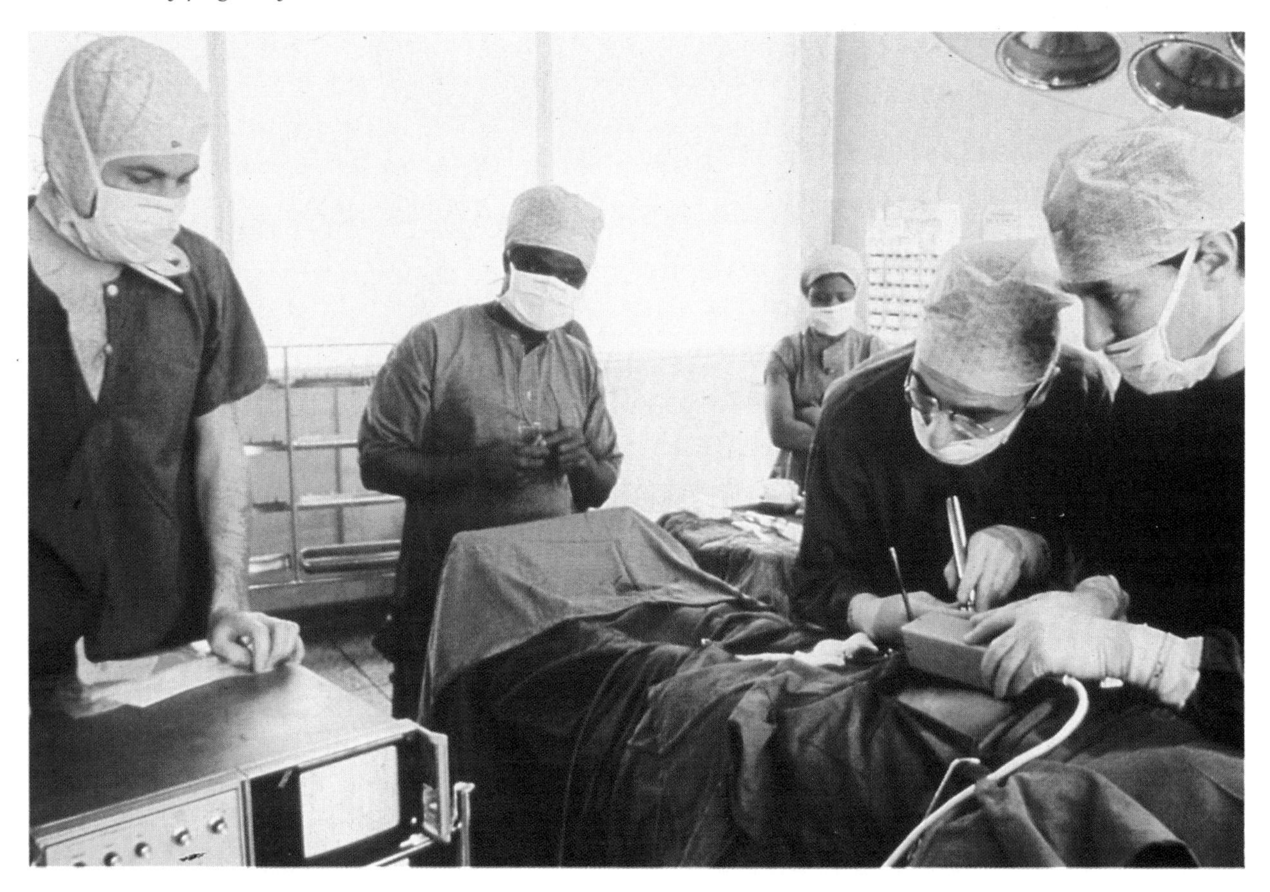

blood from the mother at seventeen weeks gestational age. The mother can then be scanned using ultrasound and the amniotic fluid tested for an excess of the chemical alphafetoprotein (using immunoelectrophoresis). The advent of screening as well as diagnosis of developmental malformations of the central nervous system has made great strides by using both immunological techniques and ultrasound.

It is inevitable that in dealing with the ethics or morality of prenatal diagnosis, screening and possible selective abortion (where permissible) for genetic abnormality, attention is concentrated on the possible undesirable aspects. It is no mean achievement to diagnose a fetus with Tay-Sachs disease or to tell a woman who becomes pregnant at age forty that, for the great majority of cases, the child will not have Down's syndrome but if it does she has the opportunity of selective abortion. The real and obvious benefits are that prenatal diagnosis and selective abortion will become widely available, and the facilities for carrying it out will increasingly expand.

There is still a large number of questions that remain to be resolved. The question about the degree of severity versus the postnatal treatment, and the question of unexpected findings, will be discussed further as the scope of the procedures widens. In addition there are good reasons for appraising the long-term social consequences which will result as the procedure becomes extended. However, when applied with care and discrimination, the benefits that prenatal diagnosis can produce are quite clear, and there remains no doubt that these procedures represent a valuable advance in the practice of medicine.

The Future for Genetic Counseling

It is sometimes said that selective abortion after intrauterine diagnosis is an interim measure, and

Certain blood disorders detected by prenatal diagnosis can be treated before birth by an exchange transfusion, in which the fetal blood is completely changed using a tube through the mother's abdomen.

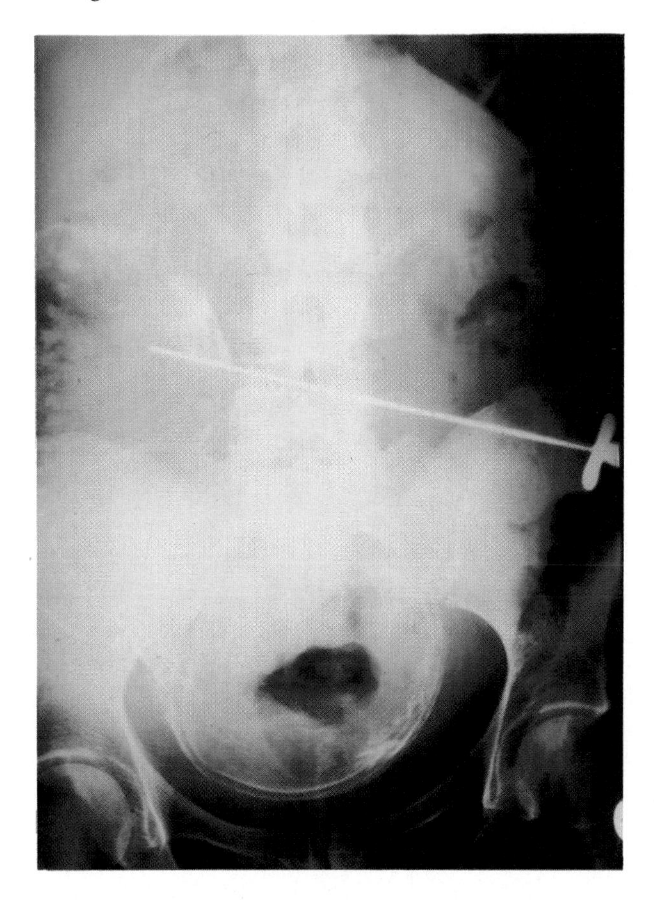

that in the future it will be possible to treat birth defects and genetic diseases either pre- or postnatally. This view is probably unrealistic. Effective treatment for a complex condition such as Down's syndrome and similar structural defects is difficult to imagine. For this reason, abortion for genetic defects discovered by prenatal diagnosis is probably here to stay for a long time.

Genetic counseling can be seen as a voluntary negative eugenics program which provides genetic screening tests, along with education and appropriate prevention such as AID (artificial insemination by donor) or contraception advice for those couples concerned about the possibility of having an abnormal child. In the past, preconceptional screening and counseling centered largely around the possibility, based on a patient's family history, that a particular disorder might occur. These predictions may now be altered in some cases through a simple blood test, which makes it possible to determine whether an individual possesses the recessive trait for certain diseases such as sickle-cell anemia, thalassemia, or Tay-Sachs disease.

Based on test results, genetic counselors can give prospective parents more precise risk figures on the likelihood of the birth of an affected child. Taken together with postconceptional screening, as with amniocentesis, a more definitive estimate of the risk of some genetic disorders can be given via counseling. If genetic counseling is neglected, there is little point to the large-scale screening programs that are currently being planned or implemented. In fact, such screening programs may do more harm than good if not accompanied by counseling or if the counseling given is inadequate or misleading. Various follow-up groups exist to aid those suffering from genetic disorders. The importance of these groups cannot be overemphasized, because they also contribute much to research.

In the prevention of genetic disorders, the use of AID is already an accepted technique in the situation where both parents are heterozygotes for the same recessive gene. As an alternative, couples may prefer the intrauterine fostering of a blastocyst produced by *in vitro* fertilization where both sperm and ovum are donated. Conversely, if a mother has a genetic condition likely to cause intrauterine damage to a fetus, surrogate motherhood — that is, the transfer of the blastocyst to a volunteer host mother — may become acceptable practice, but within strict legal and ethical criteria.

Many people wonder what burden would result from the successful treatment of genetic disorders, especially those now untreatable that effect the survival, fertility and fitness of the patient. If there were no more selection against these disorders, or if treatment for all of them became effective immediately so that individuals having them would be restored to full fitness or fertility, it has been estimated that it would take one generation to double the frequency of dominant conditions, four generations for X-linked disorders, fifty for autosomal recessive disorders and between fourteen and twenty-three generations for the multifactorial major malformations. In the meantime medical practice will no doubt learn how to handle things differently.

Chapter 7

Changing the Blueprint

As knowledge of genetics increases, so do the possibilities of somehow shaping the characteristics of future members of the human race. The expansions of the science of genetics into genetic counseling, part of which depends on techniques for predicting and detecting abnormal embryos, and into the new capabilities of genetic engineering, which bring about the possibility of correcting gene defects in both embryo and adult, may help to create the society of the future.

Attempts to modify or perfect the natural constitution of the human race — described by the general term eugenics — have been made since the earliest times. In a simple, but gruesome, form, so-called "negative" eugenics was practiced by primitive societies in which undesired offspring were killed or so neglected that they died before becoming old enough to reproduce. The victims were often female children — less valued than male children in some societies — although evident malformation, or even apparent weakness or small size, were also possible qualifications for such a brutal death. The practice was common among the ancient Greeks, as described in the legendary story of King Oedipus — who was born with a club foot.

Another example of a harsh attempt at genetic control comes from the Middle Ages, when epileptic males in Scotland were castrated to prevent their undesirable medical condition being perpetuated within the population. Similar policies, indeed, were still to be found in parts of the developed world even as late as in the 1960s.

The atrocities committed in the name of medical "experiments" by some Nazi doctors in Germany before and during World War II have been cited as perhaps the most abhorrent of eugenic practices. By themselves they have been largely responsible for making the concept of eugenics objectionable to many people. Policies such as the sexual sterilization of the feeble-minded, or of persistent

Science fiction is fascinating — and the best of it has some basis in scientific fact. In the film Run to the Stars, *the world is populated by a society of grand technological achievement and individuals very different from ourselves. If this world is a projection of our own, then genetic engineers will probably play a major part in its development.*

Although presented here as a joke, the increasing capability of genetic engineers to manipulate and modify organisms, however well intentioned, remains a cause for concern in the minds of some people.

criminals, are easily equated with the murder of those Adolf Hitler deemed unfit to be part of the German race he planned, and to most people are in consequence deeply offensive.

The term eugenics actually has a far more general, less sinister, meaning than such horrors suggest. It was coined in 1883 by the English scientist Sir Francis Galton, cousin of the great naturalist Charles Darwin. Galton — who is often referred to as the father of eugenics — was keen to advance the idea of improving human stock by encouraging mating only between what he saw as the more suitable social classes and races. From his studies of eminent families he concluded that it would be possible to produce a highly gifted race by judicious marriages over several generations. His ideas encouraged eugenic movements in the United States and Europe, and disciples of Galton thought it reasonable that geneticists should be working toward improvement of the human species by selective breeding. The fact remains, however, that until recent times scientific eugenics was nowhere near becoming a practical possibility, and even now cannot become such until the mode by which all human characteristics are transmitted is fully understood.

Many ethical problems are raised by the question of how eugenic practices might be administered. Because of these, it seems that the only way eugenics could be made acceptable on a wide scale would be if society — which means all its members individually, and not merely a majority — could decide and select those qualities it deems most desirable, such as perhaps intelligence, beauty, or physical strength. Even such a simple list illustrates the problem: if *everyone* was beautiful, what would be the point in beauty (. . . which, we are told, lies more in the eyes of the beholder than in a definable standard anyway, and so is not measurable).

Apart from ethical and subjective dilemmas there would also be practical problems. To realize this form of eugenics it would be necessary to be sure that there were no recessive, unexpected or unwanted characteristics present in any of the mating population. If these existed, some might combine with other recessive characteristics to form "undesirable" traits.

Furthermore, the crucial problem in even imagining, let alone trying to achieve, such a "perfect" society is that the very qualities that everyone could agree to admire are highly complex genetically, being the result of not one but many genes. The characteristics that can be linked to a very few simple genes are among the most simple, physically: blue or brown eyes, straight or curly hair, dark or light skin. And apart from their genetic complexity, most desired qualities in people are also strongly influenced by diet, upbringing, education and environment.

Eugenics and the Law

Eugenic ideas are sometimes given legal status. For example, many American states forbid people to marry their ancestors or descendants — they prohibit any lineal marriages between blood relations. There are also states that do not allow colateral consanguinity, which describes marriage between people descended from a common ancestor but not from each other, such as cousins. Some states, such as Georgia, forbid nephew-aunt marriages but do not disallow those between uncle and niece.

Only a generation ago a number of American states prohibited marriages between two people of different races. For example, blacks were not allowed to marry whites in Virginia until the law concerning this was successfully challenged in 1966. Richard Loving, a white man, married Mildred Jeter, a black woman, and moved from Washington D.C. to Virginia, where he was promptly convicted of violating the law against interracial marriage. The case was put before the Supreme Court of the United States, which overturned the conviction. In South Africa, "mixed" marriages between blacks and whites became legal only in 1985.

There are also laws that prevent marriage involving the mentally retarded, because they may be too incompetent to appreciate the responsibilities of the marriage contract, or to protect the infirm from others who might abuse them in a marital relationship, or as a way of inhibiting the procreation of children whose mental defectiveness may be inherited. These eugenic laws protect mentally retarded individuals against abuse but also cushion society against the responsibility of looking after their descendants. This latter aspect is abhorrent to those who believe in the sanctity of life above all else. However, the principle of eugenics tends to favor society more than the individual.

Through a more widespread understanding of the causes and effects of physically disabling disorders, most people feel less threatened or embarrassed by the presence of the handicapped in our society. In the future, genetic counseling and genetic manipulation may dramatically reduce the incidence of such disorders.

Predicting the Future

In some countries, genetic counseling enables people to administer their own form of eugenics — elective abortion in response to the knowledge that a growing fetus will be born as a defective child. Developing techniques to detect abnormal fetuses *in utero* will also change the shape of society. Amniocentesis, developed in the early 1960s, has proved a major tool in identifying abnormalities. The technique, which involves the removal of a sample of amniotic fluid and its subsequent analysis, can detect various chromosomal disorders, such as Down's syndrome. In consequence, the number of Down's babies has been considerably reduced, particularly among the children of older women who are more likely to give birth to them. Other conditions such as spina bifida and severe metabolic disorders can also be identified using this method, which is described in detail in Chapter 6.

Ultrasound (sonography) in the 1970s followed

Beauty and intelligence can both be inherited. Comparing them, the dramatist George Bernard Shaw said: "Beauty is all very well at first sight; whoever looks at it when it has been in the house three days?"

amniocentesis as a valuable method of identifying fetal abnormalities, particularly those of the nervous system, kidneys, gastrointestinal tract, limbs or heart. Ultrasound provides an image like a television picture of the fetus; experienced operators can use it to identify abnormalities and thus warn the mother so she can consider the possibility of abortion.

Ever more sophisticated techniques are continually being developed to enable doctors to know whether an unborn child is likely to suffer from such diseases as Huntington's chorea (dementia with involuntary muscular twitches) or other inherited disorders. Gene probes and other genetic techniques promise to make it possible to identify a genetic predisposition to breast or colon cancer, diabetes, schizophrenia, or Alzheimer's disease (premature senility).

The danger of such predictions is that some parents might overreact if they know their unborn child is likely to have a predisposition to a feared disease, for example diabetes, and decide to have an abortion. At present the circumstances that might make such predictions possible with some degree of certainty are still remote, but the time has nearly arrived when a woman could have a number of her ova fertilized in a dish and have each one tested to distinguish which has the most favorable set of characteristics. A more likely outcome of the use of gene analysis to predict the future, however, is that it could become possible for people predisposed genetically toward a condition such as heart disease to be made aware of this and so to be able to choose a lifestyle that reduces other causative factors such as a fatty diet or smoking.

Genes and Genius

Some people have expressed the view that a slow decline in the overall intelligence of society can be expected because those with low intelligence tend to have larger families than those who are highly intelligent. This carries the implication that those of low intelligence, and especially those who are mentally retarded, should be prevented from having children, presumably by means of sexual sterilization. Such statistical differences in family size can also be explained by the increased educational and job opportunities for intelligent people, who are in consequence more likely to be in occupational groups that practice some form of family limitation.

The facts disprove the large-family-low-intelligence theory, however. In Western countries, parents with the largest families have average or even above-average intelligence, and the severely mentally retarded are usually subfertile or even infertile. Also, statistical analysis of variation in intelligence in any population shows that variation occurs in the middle range of intelligence, not in high- or low-intelligence groups. At present there is no evidence to prove that natural human intelligence is declining, and there is thus no justification for applying negative eugenics by the sterilization of the mentally retarded.

Various suggestions have been made from time to time regarding ways to produce a population with a greater proportion of outstanding people. One possibility, for instance, might be to allow tax relief or give monetary allowances to people whose attributes society wishes to propagate and thus to provide adequate facilities for them to have many children. But questions immediately arise. Would the "desirable" parental attributes actually be present in their children? Who would decide what characteristics and which people qualify for this special treatment? And, furthermore, can anyone be sure that the sort of qualities that are preferred today will be those preferred by generations of the future?

In the last-mentioned context, it is worth remembering that some people in the past whom we revere today were far from physically or mentally perfect. Dostoevsky and Julius Caesar were epileptics. Van Gogh was psychotic. Byron and Toulouse-Lautrec were physically deformed. According to the 1925 Eugenics Society, the characteristics of all these people would have been undesirable.

Today, the only real eugenics is practiced by the individual who chooses whether or not to bear children, and by the laws that prevent consanguineous marriages, which genetic analysis shows can lead to abnormal offspring. We are not yet ready for the Brave New World of Aldous Huxley, in which everyone is given a particular predetermined set of genes, thereby controlling their

There is a remarkably long list of famous people who have been physically or behaviorally imperfect, yet their words and deeds are remembered long after their death. The English poet and satirist Lord Byron, whose accomplishments brought him lasting fame, had the misfortune to be born with a deformity in one of his feet.

The help of alchemists was once sought by people wishing to have children of a particular sex, for social and economic reasons which have largely disappeared. Many parents still would like to be able to choose.

Artificial insemination enables farmers to produce animals from parents located in different parts of the world. With unlimited access to banks of ova and sperm, farmers could produce animals "to order."

appearance, their intelligence and thus their station in life.

Huxley's fiction is the total antithesis of the American dream that everyone can be a success. Because of modern developments in genetics, however, we are far closer to the possibility of realizing the genetic practices described in *Brave New World* than the world was in 1932 when Huxley wrote the novel. In fact, another famous writer's words describe, in a far more optimistic way, the situation, even the ideal, toward which medicine is moving, thanks to recent genetic research. The words are those of Hippocrates, who proclaimed that the aim of medicine should be to "Declare the past, diagnose the present, and foretell the future."

A Boy or a Girl?

Advances in genetic testing have begun to make it possible to answer this age-old question with certainty long before the moment of birth. Indeed, there are even those who are trying to predetermine the sex of their children by attempting to use one or other of a wide variety of sex selection techniques before conception.

In the Middle Ages, one method used to influence the chances of giving birth to a boy involved drinking an alchemist's preparation of wine and lion's blood, then having intercourse under a full moon while an abbott prayed for the couple. Today many Americans follow the theory supported by some medical as well as nonmedical authorities, that sperm are "male" or "female" and that although "male" sperm swim faster than "female" sperm, the "female" sperm can survive longer in a woman's genital tract. Ovulation occurs on about the fourteenth day of a twenty-eight-day cycle and the theory states that if there is intercourse on the twelfth day only, "female" sperm alone will survive to reach the egg. But if sexual intercourse takes place only on the fourteenth day, so the theory goes, the faster swimming "male" sperm will reach the egg first. Recent

Feeding the world's growing population could depend on genetic engineers being able to improve the yields and tolerances of food plants. Suitable candidates include the fast-growing, oil-seed bearing sunflower.

evidence from New Zealand, however, proposes the complete opposite. There the considered advice to prospective parents is: make love early for a boy and late for a girl.

Another possibility under investigation by researchers is that the vagina's acidity affects the outcome. It is thought that "male" sperm are more active in acidic conditions, whereas "female" sperm are more active in an alkaline environment. Following their belief in this hypothesis, some women desperate for a boy or girl have even gone so far as to use a douche of vinegar (acidic) or bicarbonate of soda (alkaline) to influence their hoped-for child's sex.

Yet another theory is held by the French gynecologist Dr François Papa, who believes that by manipulating her diet a woman can influence the sex of her child. Quantities of salt are recommended to produce a boy, although oysters, shrimps and wholemeal bread are out. For a girl, a pint of milk a day is needed whereas coffee, tea, alcoholic drinks, mushrooms, salad tomatoes and avocados are forbidden.

Finally, perhaps the only genuinely scientific approach among these options is being evaluated in Japan. Japanese scientists think the answer may lie in the separation of sperm into those carrying the X chromosome, which combines with the egg to produce a girl, and those containing the Y chromosome which results in a boy. If experiments to inseminate animals according to this principle and to predict the gender of the offspring are successful, then the technique could in the future be applied to humans.

There is still no conclusive proof of whether any of these methods of influencing a child's sex actually works — but perhaps it is just as well we do not know for sure. Otherwise the natural balance could be disrupted and people could one day find themselves in a world where there were too few women to give birth to the next generation, or discover that there were only a handful of men available as potential fathers.

If we were indeed able to influence the sex of our children, would many of us really want to? The answer seems to be "Yes." A recent survey in The United States showed that 39 per cent of women were in favor of predetermining the sex of their children. Another study showed that two-thirds of couples had a preference for the sex for their firstborn. Of these, 90 per cent wanted a boy. In a society where it is boys who carry the family name forward it is easy to understand the desire to have at least one boy in the family. And parents with only boys or only girls might be grateful for a technique that could alter the pattern. A good medical argument for choosing girls rather than boys applies in cases where parents want to avoid sex-linked genetic disorders such as hemophilia and one type of muscular dystrophy, where there is a chance that a boy will inherit the disorder and be affected but not a girl.

Although influencing the sex of a child before conception still remains a lottery, discovering a child's sex once it has been conceived is now relatively straightforward. The techniques are similar to those used in prenatal diagnosis, and include amniocentesis, ultrasound (sonography) and even tissue sampling as microsurgical methods are refined and improved.

The Technology of Life

Genetics has already made contributions of incalculable value to the health and prosperity of the world through research and the discovery or invention of practical solutions in three principal areas: agriculture, animal farming and medicine. The prospects for the future are correspondingly bright. Biotechnology — which is the general term

Crown gall disease (left), *a type of plant cancer, is caused by harmful microbes, such as* Agrobacterium tumefaciens (right). *Yet the same bacterium is a valuable agent for transferring genes between species*

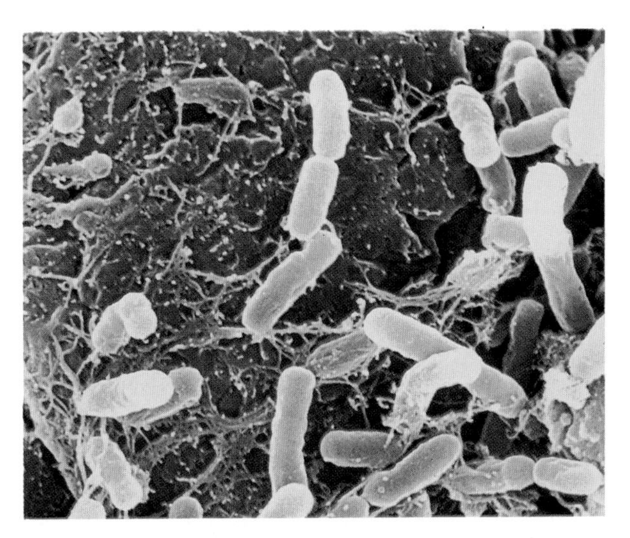

Current methods of increasing the size of farm animals is limited by the animals' growth-regulating hormones. But continued research could lead to genetically engineered supercattle for food.

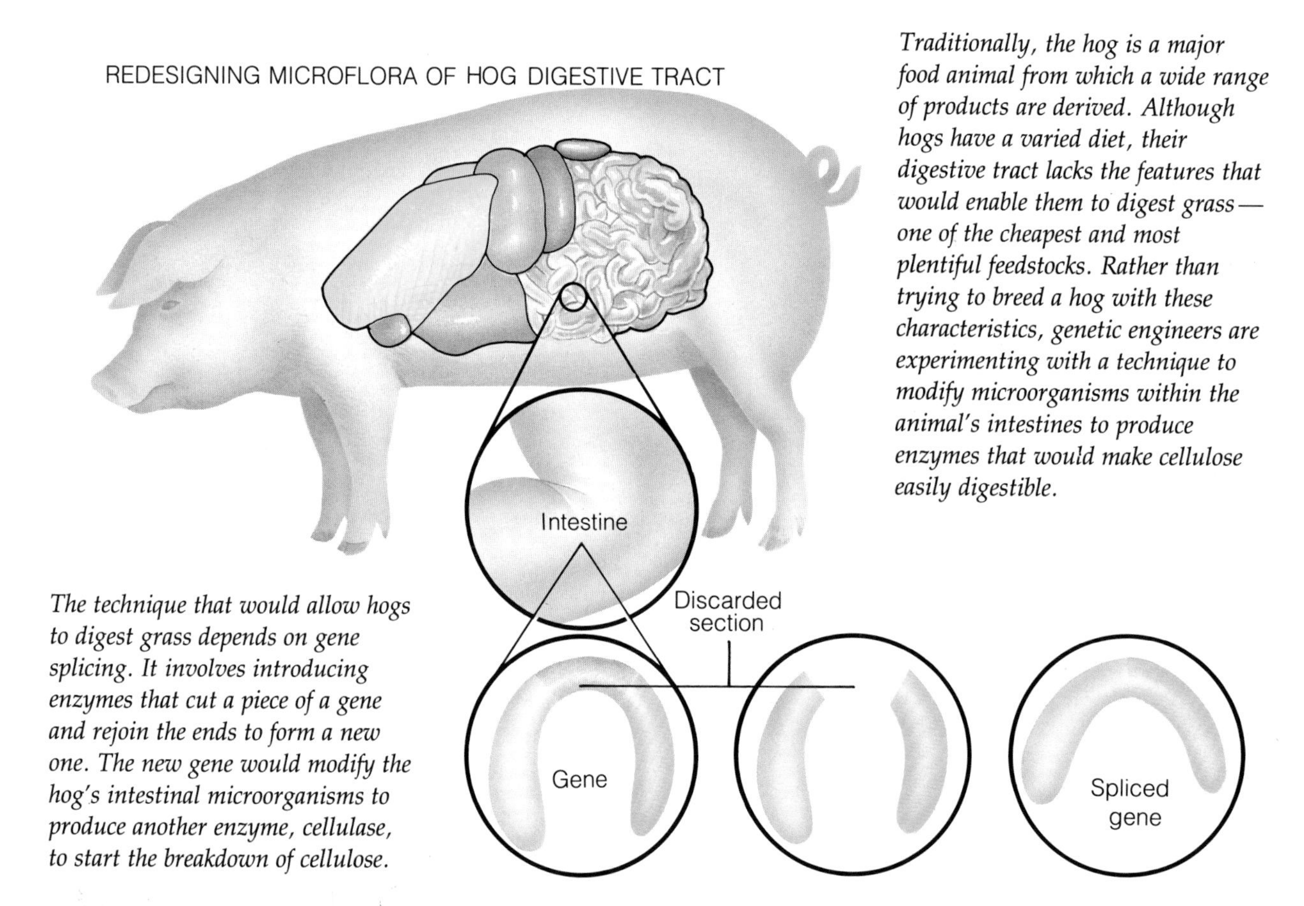

Traditionally, the hog is a major food animal from which a wide range of products are derived. Although hogs have a varied diet, their digestive tract lacks the features that would enable them to digest grass — one of the cheapest and most plentiful feedstocks. Rather than trying to breed a hog with these characteristics, genetic engineers are experimenting with a technique to modify microorganisms within the animal's intestines to produce enzymes that would make cellulose easily digestible.

The technique that would allow hogs to digest grass depends on gene splicing. It involves introducing enzymes that cut a piece of a gene and rejoin the ends to form a new one. The new gene would modify the hog's intestinal microorganisms to produce another enzyme, cellulase, to start the breakdown of cellulose.

If gene-splicing experiments to modify the intestines of hogs to digest cellulose are successful, the animals might develop a taste for grass and be put out to pasture like cattle and sheep.

for the wide range of such activities, many involving genetics in a central role — is predicted to become one of the world's foremost industries by the year 2000, with billions of dollars' worth of genetically-engineered products produced and sold every year.

Already there has been massive investment in the industry and companies are now beginning to reap the rewards of years of painstaking research. Genetic engineering promises to allow plant breeders to insert foreign genes into plants, which will be able to produce crops that yield more calories and protein, resist major crop diseases, grow during droughts or in salt-laden soils, withstand extremes of cold and heat, be immune to the herbicides used in pest control, and flourish without the addition of artificial fertilizers.

One technique that is leading to these new genetic horizons and so could help combat famine around the world involves using microbes to transfer useful genes into plants. A major breakthrough in the development of this technique came in 1981 when American scientists announced that they had transferred a protein gene from a bean seed to the cell of a sunflower, a member of another plant family, to produce an entirely new variety of plant which they called a "sunbean."

Sheep shearing is an expensive, laborious task, even with modern tools. Sheep could perhaps be genetically modified so that they are self-shearing, so that the full-grown fleece dropped off.

This gene was not only stable in the host but also produced messenger RNA, one of the major steps toward the actual expression of the gene. The transfer was achieved by inserting the protein gene into a Ti (tumor-inducing) plasmid from the bacterium *Agrobacterium tumefaciens*, which causes crown gall disease — a form of cancer — in plants, and then infecting sunflower cells with the bacteria carrying the modified plasmid. The gene from the plasmid was subsequently accepted by the sunflower plant into its own genome.

The technique demonstrated that plants could accept genes carried on bacterial plasmids. Furthermore, it showed that it was possible, by insertion of the desired gene next to a specific site on the *A. tumefaciens* plasmid, to actually get the gene to be expressed in the new host. This was demonstrated by using an antibiotic marker gene and then searching for antibiotic-resistant plant cells. Fortunately, some forms of *A. tumefaciens* were found carrying plasmids which did not lead to crown gall disease in the host plant.

The discovery that the genes introduced by the plasmid were passed on to the next generation added to the importance — and the excitement — of the results of the experiment. This demonstration of the mechanism for the transfer of genes into plant cells opened the way for the transfer of almost any desirable gene. Viruses too are being considered as vectors for transferring genes when a bacterium will not infect a plant to which a specific gene is to be transferred.

The next step in this agricultural — and genetic — adventure is to characterize more genes from thousands of plant species which may carry up to 30,000 different genes, only a few of which have been studied in detail. Scientists in the future should be able to transfer genes into almost any plant to provide it with any combination of specifically desired properties.

Genetic engineering will also bring advantages in animal breeding, leading for instance to improved meat production. Recent experiments in the United States have shown that if rat growth-hormone genes are injected into fertilized mouse eggs, which are subsequently implanted in foster-mother mice, the result is the production of "super-mice" which grow to nearly twice the normal size. Some of these mice have passed on the implanted genes to their own litters. As Dr Ralph Brinster, one of the two scientists involved in this work, put it, "If we can make bigger mice, we can make bigger cows."

Herein lies the future: if scientists can create bigger cows or pigs, then more meat can be supplied to an increasingly hungry world. Animal breeders will also benefit if scientists can find genes that confer disease resistance or reproductive prowess, and can clone these and then insert them into germ-cells or early-stage embryos. Such research opens up another opportunity: the use of "genetic farming" to produce larger quantities of medically useful products in the same way that "super-mice" produced increased amounts of growth hormone.

The largest constituent of plant cells is cellulose, which some animals cannot digest. Biotechnologists are currently investigating ways to change this, however. Their intention is to redesign the microflora of a hog's digestive tract, with the aid of recent research into uses of recombinant DNA, to enable hogs to produce the enzyme cellulase, which would allow them to extract nourishment from cellulose. Because a hog's digestive system is very similar to that of humans (and because humankind therefore competes for a number of

To manufacture interferon — the protective protein produced within cells — genetic engineers break open plasmid DNA from the Escherichia coli *bacterium, found in large numbers in the human intestine.*

A culture of E. coli *with suitable plasmids is grown, precipitated and treated with enzymes to release the plasmid DNA. This is separated and further treated with enzymes to make it receptive to interferon cDNA.*

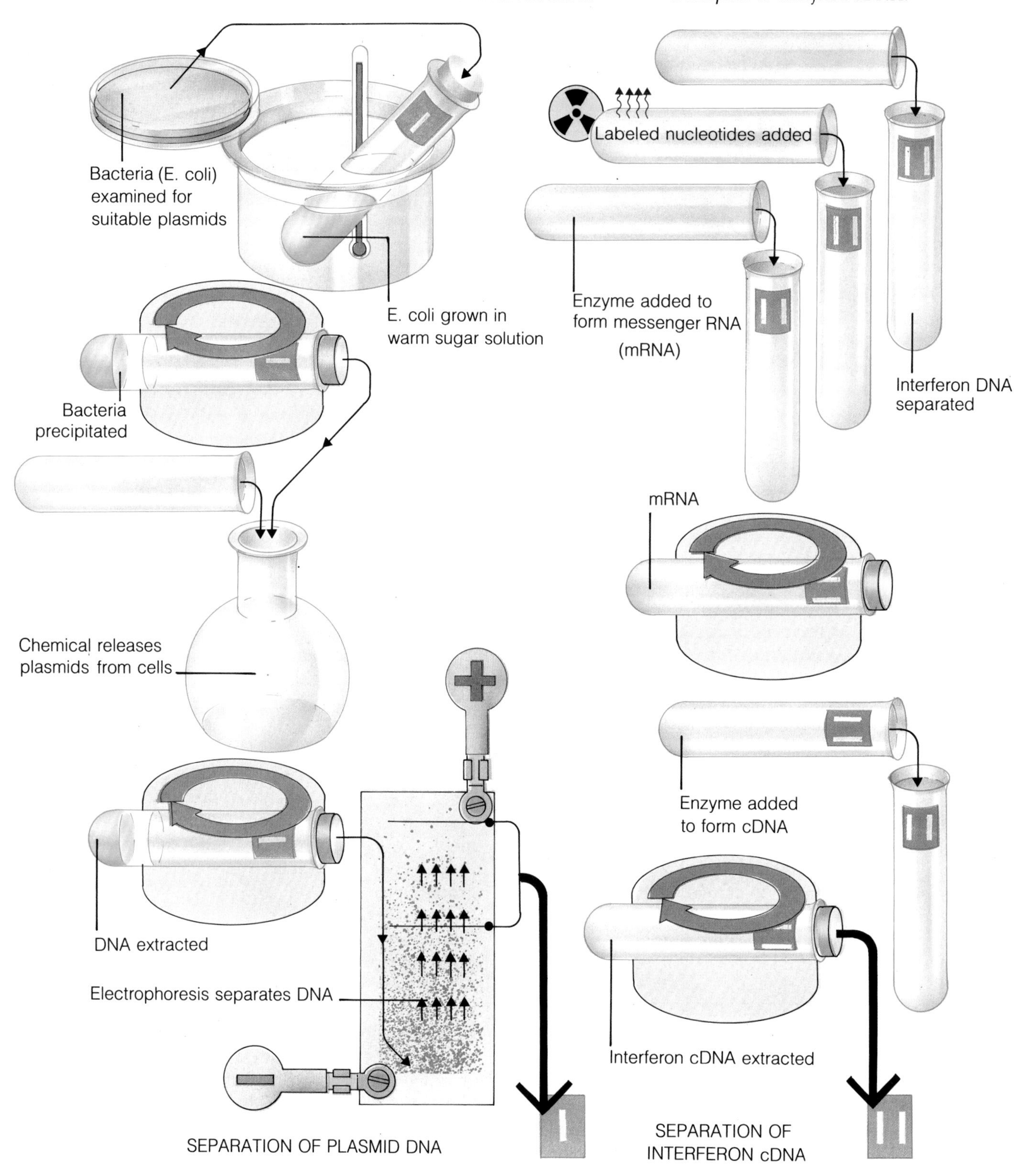

Once the plasmid and interferon DNA have been prepared, they are combined and tested to see if they will infest a new E. coli *culture. If they do, RNA is extracted to produce interferon in frog's eggs.*

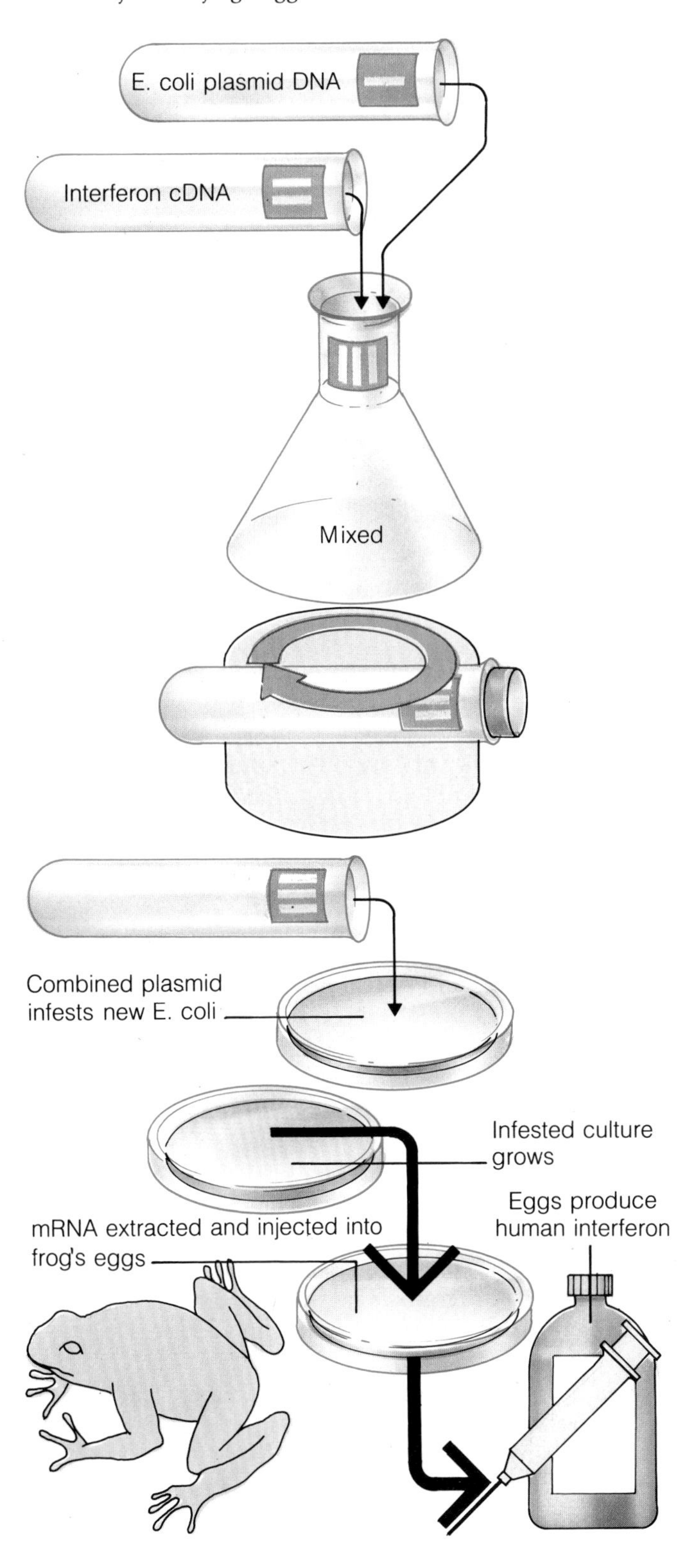

basic foods that farmers produce), it would be a considerable economic advantage to farmers and to world food supply if this biotechnological theory were to become reality.

Another economic boost for farmers could come from the cloning of the gene for a protein which, when it is injected into Merino sheep, causes the fleece to fall off. This "self-shearing sheep" protein can be extracted in small amounts from the salivary glands of male mice; by cloning the gene, there would be plenty to sell to sheep farmers at an acceptable price.

In the world of fish farming, researchers at the University of Southampton, England, are hoping to use gene technology to produce "super-trout" by incorporating genes from mice and frogs. It is hoped that these genes will produce fish which grow larger and faster, and flourish in conditions which prove lethal to normal trout. Chinese scientists are working on the genetic manipulation of carp to produce similarly "super" carp. More ominous, as far as the Chinese are concerned, is the possibility that the gene for the protein that makes silk will soon be cloned by genetic engineers. This would make it possible to create an industrial process that could make redundant the ten million farmers in China who earn their living producing silk using silkworms.

In medicine one of the most important applications of genetic engineering is in artificial hormone production. Conventionally, to obtain supplies of a hormone such as insulin, which is used to treat diabetes, it is necessary to extract it from the pancreas of an animal, traditionally a hog. Approximately one per cent of diabetics are allergic to porcine insulin, however. In 1982, in an attempt to solve this problem, the Danish company Novo Industri developed a technique to remove the end of one of the two chains which make up the porcine insulin molecule and replace it with the human equivalent. This solved the problem, biochemically. More recently, the American biotechnology company Genentech has genetically engineered bacteria to produce two amino-acid chains which can be linked to produce human insulin, thereby offering an alternative solution.

The human growth hormone, which was previously obtained from human cadavers (a

The treatment of cancer with modified genes requires a sample of bone marrow, from which DNA is extracted. The genes are separated and treated with enzymes, then reinfused into the patient.

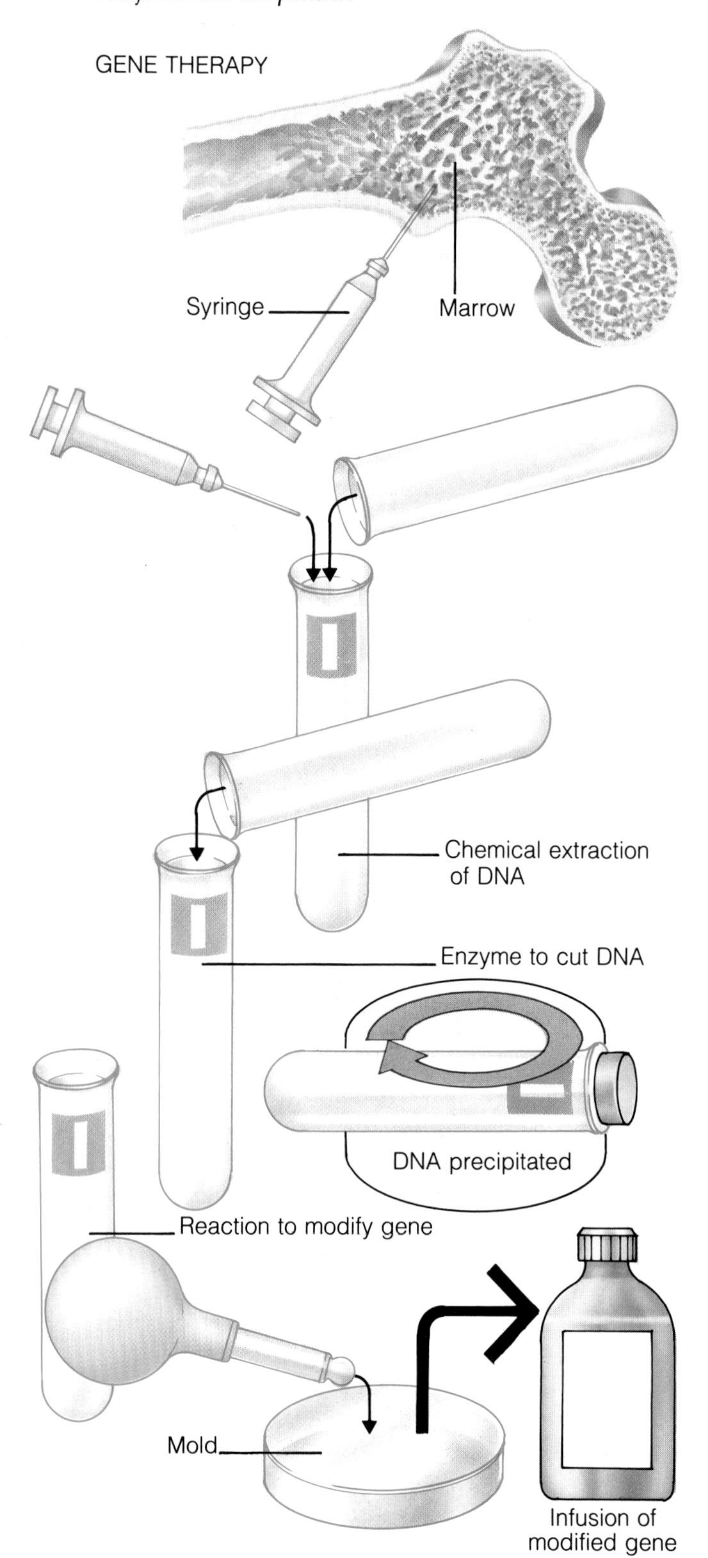

technique recently banned by the FDA), has also been produced using bacteria. The availability of a new copious source of human growth hormone raises important ethical questions, however. Should a child of short parents be given growth hormone to make him or her taller? Because physical size is also related to diet, is there a danger that use of the hormone might create artificially and unhealthily tall children? And what is the desirable height — average, or taller than average — assuming that the parents believe tallness to be desirable because of associated social, athletic or career advantages?

Genetic engineers are also currently trying to produce other hormones such as parathyroid hormone, which would be useful in the treatment of osteoporosis; nerve growth factor, to help restore damaged nerves after surgery; and erythropoietin, a hormone that can be used to help regulate blood-cell development.

A booming area of genetic engineering involves "protein engineering" to create enzymes which are more effective and efficient than those currently available. Scientists at London's Imperial College of Science and Technology, and at the Laboratory of Molecular Biology on the outskirts of Cambridge, England, have had success manipulating the gene coding of two specific amino acids in an enzyme's structure. They produced two new enzymes, one of which was twice as effective as the original enzyme; the other was twenty-five times more efficient than the original. Although their products were not commercially viable, genetic engineers hope it will soon be possible to produce powerful enzymes which can be used for specific medical and industrial applications.

Genetic engineering also seems to be on the brink of a revolution in vaccine technology, the aim being to speed up the appearance of new vaccines and to reduce the risks that are sometimes involved in immunization programs. One important advantage of using microorganisms to produce a vaccine is the relatively low cost.

It is also easier to overcome the public fear of serum-transmitted diseases if a vaccine is produced without using human blood plasma. Much of the anxiety about the acquired immune deficiency syndrome (AIDS) is generated because this disease

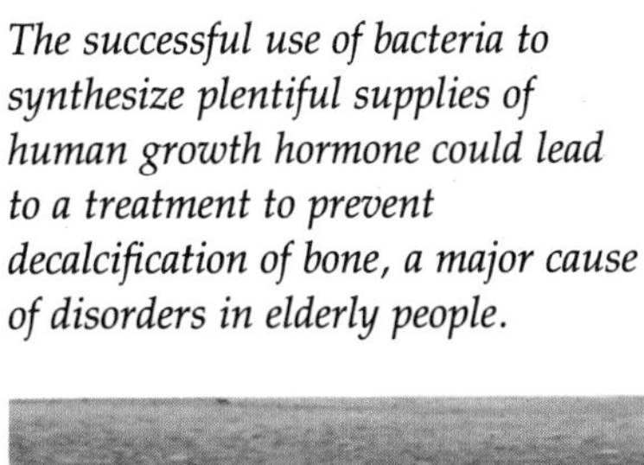

The successful use of bacteria to synthesize plentiful supplies of human growth hormone could lead to a treatment to prevent decalcification of bone, a major cause of disorders in elderly people.

can be transmitted in human blood plasma, which is also used in the conventional production of hepatitis-B vaccine. Hepatitis-B is a chronic disease transmitted by contact with infected blood, saliva or semen, and there are thought to be around 150,000 new cases a year in the United States. Once genetically-engineered hepatitis-B vaccine has passed clinical trials it could be used more commonly than at present, when authorities can immunize only those at high risk of contracting the disorder, such as physicians, dentists, dialysis patients and others likely to come into contact with infected blood.

Many vaccines are produced by growing them in living tissues, and there have been problems associated with vaccines other than hepatitis-B, although not all have affected humans. For instance, some outbreaks of foot-and-mouth disease affecting livestock have been traced to contaminated vaccines. In the United States, new veterinary products may be approved after only two years' testing, whereas five to seven years are necessary for acceptance of a human vaccine. It is therefore likely that genetically-engineered vaccines will be used for animals in the near future.

Research is also being conducted into finding or making vaccines for diseases such as genital herpes and AIDS, which cannot be prevented at present.

Genetically-engineered antibiotics are another prospect. Some of the body's own natural defensive cells, such as phagocytes, are able to absorb antibiotics, which enhances their effectiveness against bacteria. Another type of cooperation between defensive cells and antibiotics is the way in which an antibiotic drug changes the shape of bacteria, making them more vulnerable to attack by the body's immune system. With these points in mind, scientists are trying to produce antibiotics which not only kill invading pathogens but also help boost the body's immune system.

The genetic engineering of human beings to produce people with a whole range of predetermined characteristics in the model of Huxley's *Brave New World*, and in other science fiction, is still

The problem of unwanted children has not been solved, despite the availability of contraceptives. An old print shows abortionists disposing of the evidence (they were subsequently arrested and brought to trial).

The well-established technique of in vitro *fertilization enables women with blocked Fallopian tubes to bear children. Surplus embryos from this procedure could provide graft tissue to promote growth in adult organs.*

too far from reality to bear serious consideration. There are, however, grounds for concern about some of the effects of genetic engineering. For example, if super-bacteria which are resistant to antibiotics were accidentally released into the environment, the effects would be the same as germ warfare.

One incredible prospect concerns *Escherichi coli*, a bacterium found naturally in the human intestine, which is one of the most commonly used genetic-engineering vehicles. It is possible that this bacterium could be manipulated to produce industrial alcohol. If these altered, alcohol-producing bacteria escaped they might become established in human intestinal tracts, where they would continue to produce alcohol, with predictable and alarming results! Because of such possibilities, many countries have special bodies to police the biotechnology industry and guidelines exist to ensure safety is maintained and that members of the general public are protected.

The Future: Hopes and Fears

Although scientists are far away from the full-scale genetic engineering of humans, they may soon be able to use it in a limited way, for instance to perform gene therapy. For example, bone marrow cells could be extracted from a patient with a genetic blood disease such as thalassemia, and the genes modified to correct the defect. Such a process was attempted in 1980 by a professor of medicine at the University of California, but was unsuccessful.

The experiment was performed without the benefit of all the research on animals that has been conducted since. From this, so much knowledge is being amassed that before the end of the century it may be possible to use gene therapy to compensate for defective genes in some of the 3,000 or more known genetic diseases, from mental retardation to blood disorders. Because the resulting changes would not be passed on to another generation, but would affect only the individual treated, gene therapy is likely to avoid most of the ethical problems that arise with more fundamental genetic manipulation.

Coming closer again to controversial issues is the possibility of genetic alteration in humans at the stage of the fertilized egg. If a fertilized egg is recognized as being defective, it may one day be possible to replace certain genes to correct the defect using "gene surgery."

The possibility of "gene surgery" arises from the practice of *in vitro* fertilization, which began in 1978 after the pioneering work of the now world-famous British team of obstetrician Patrick Steptoe and reproductive physiologist Robert Edwards. The technique was finally proved successful with the birth of the world's first "test-tube baby," Louise Brown, born July 25, 1978.

In producing an *in vitro* fertilized egg, several eggs extracted from the woman are mixed with sperm. This can result in the production of a number of embryos, one of which can be reintroduced to the mother's uterus, where it continues to develop through the normal stages of embryonic and fetal life.

Robert Edwards and others who share his views now want to be able to carry out research on the "spare" embryos that are not implanted in the mother's womb, so that genetic engineers may one day be able to correct genetic defects by the injection of "normal" genes. This technique works in mice but scientists need to discover how it could work in humans. An intensive program of embryo research, they maintain, is the only way to find out how many copies of the gene should be introduced, where it should be inserted into the cell genome, and how it might be regulated.

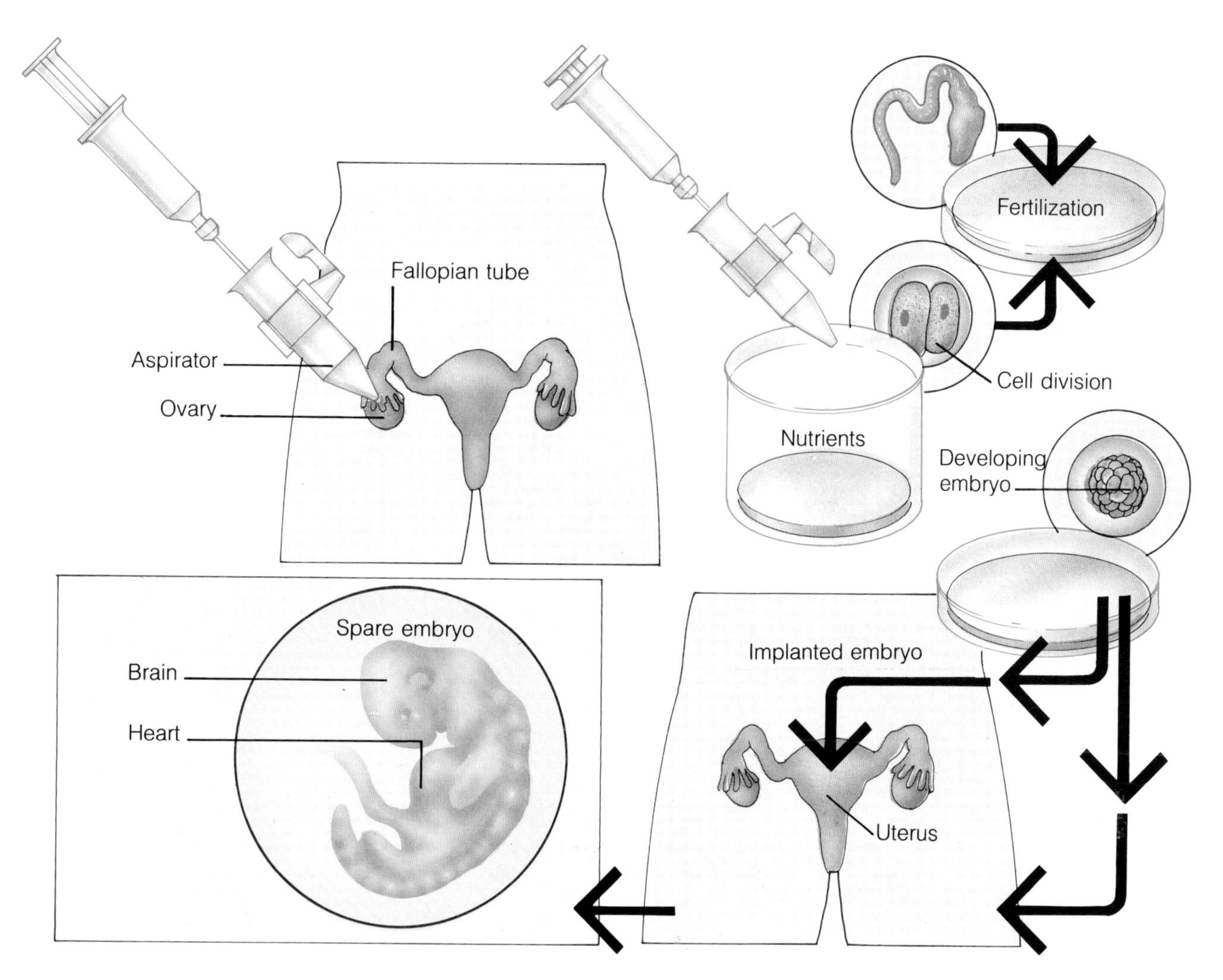

The point of conflict — and the reason such research is restricted — is that many people believe a human being is created from the moment an ovum becomes fertilized, and that in consequence such experiments would in fact be the same as human vivisection (with all the attendant legal complexities). The idea that such research might also create monsters (like the one of Dr Frankenstein in Mary Shelley's novel of that name, which was published in 1818), exists only in the popular imagination, although it is necessary to recognize that it reflects a genuine fear of the dangers of tampering with nature.

Embryo research could also provide much information about how organs differentiate normally or abnormally, and about the effects of environmental agents on embryo growth and human genes. This would be of enormous value for determining which agents are potentially mutagenic to humans. Embryo experimentation could also enable scientists to study the possibility of grafting stem cells of various embryo organs into adults. There are no dividing stem cells in adult organs such as the brain or the muscles of the heart, so natural repair is limited. By grafting embryonic stem cells on to damaged tissues, repair could be encouraged and so lead to a viable treatment for a number of disorders in adults and children.

At present the necessary microsurgical techniques would be difficult to perform, but once such a practice became established, techniques would inevitably improve. The potential use of embryonic tissue for grafting into adults is highly speculative but is nevertheless an important possible benefit from the development of *in vitro* fertilization and the associated research using embryos. Society is perhaps not yet ready to allow scientists unlimited embryo experimentation. But if and when the public feels ready for it, scientists are waiting to enter a world previously only imagined in science fiction.

Society also fears that eugenic possibilities associated with cloning embryos — that is, copying the genetic makeup so as to produce an exact replica, or numerous replicas. Could this lead to a race of super beings?

Glossary

acquired immune deficiency syndrome virus disease which disrupts the body's defense against invading organisms increasing susceptibility to infections.

adaptation or **adaption** the process of change in a characteristic of body or behavior that fits an organism to its surroundings; the resultant characteristic.

adenine a base found as one of the constituents of DNA. In the double helix of DNA it is always bonded to thymine.

AID artificial insemination by donor, in which semen is taken from a man and artificially placed in a woman's reproductive tract.

AIDS acquired immune deficiency syndrome.

albinism a condition in which there is a total or partial lack of pigment in the skin, hair and eyes.

allele one of a pair of genes that can appear as alternatives in heredity, which are located on equivalent portions of chromosomes.

allelomorph allele.

Ames test a test for possible cancer-forming agents, based on the rate at which they cause mutations in bacteria.

amino acid nitrogen-containing organic acid that makes up the building blocks of proteins; there are twenty varieties of amino acids in most human tissues; every protein contains hundreds of amino acids.

amniocentesis withdrawal through the abdominal wall of a little amniotic fluid from the womb of a pregnant woman for analysis. It allows early diagnosis of some abnormalities in the fetus.

amnion the membrane around the fetus, which encloses it in a fluid-filled sac.

amniotic fluid the fluid in the amnion.

anemia a condition in which the number of red blood cells and concentration of hemoglobin in the blood is lower than normal.

antibiotic a substance, natural or synthetic, which kills bacteria.

autosomal pertaining to any of the chromosomes other than the sex chromosomes.

banding the process of dyeing chromosomes to produce bands that can be used to identify specific regions.

bases compounds with an alkaline reaction, such as the thymine and adenine that are joined in part of the structure of DNA.

capsular belonging to the outside coat of a structure or organism.

carcinogen any substance or agent that can produce a cancer.

carcinoma a type of cancer starting in the epithelial cells of the skin or membranes of internal organs.

carrier describes a person or organism in which a certain gene is present. The gene may show no effect because its expression is masked or overruled by another gene; such a masked gene may show its effect in the next generation.

catalyze to help or speed up a chemical reaction, as enzymes do.

cellulase an enzyme that mediates the breakdown of cellulose.

cellulose a large carbohydrate molecule that forms the cell wall of plants.

centromere a constricted area on a chromosome, not necessarily near the middle, that divides it into two arms.

chemotherapy the treatment of a disease by chemical methods.

chiasma a point at which material is exchanged between paired chromosomes during cell division.

chorion a membrane around the fetus outside the amnion.

chromosomes the structures within the nucleus that contain the genetic information, in the form of DNA. Chromosomes normally occur in pairs, and become visible as rodlike bodies at the time of cell division.

clone an identical copy of an animal, plant, cell or molecule produced from a single starting cell or molecule.

codominance the situation in which two or more of several possible forms of a gene are equally dominant.

colon the large intestine, the last part of the gut.

combinative generation reproduction in bacteria which includes genetic exchange between individuals.

congenital describes a condition present at birth.

conjugation a form of bacterial reproduction in which cells join and pass genetic material from one to another.

controller gene a gene that controls the action of another gene or group of genes by switching them on or off.

cross-pollination the application to the female part of a flower of pollen from a plant of a different species.

crossing-over exchange of material between a pair of chromosomes during meiosis.

cytoplasm the substance within a cell apart from the nucleus.

cytosine a base found as one of the constituents of DNA. In the double helix of DNA it is always bonded to guanine.

deletion removal of a short sequence from a DNA chain.

deoxyribonucleic acid DNA.

dialysis a process for removing substances from a solution by filtering it through a membrane. A dialysis machine may be used as an artificial kidney.

differentiation the development of specialized tissues or cells from unspecialized forerunners.

dizygous formed from two different eggs, as with nonidentical twins.

DNA a compound found in the nucleus of cells which, together with attached proteins, forms the chromosomes. A long-chain molecule, DNA contains in its structure the coded information needed for growth, replication of cells, and heredity.

dominant describing a gene or characteristic that always manifests itself, even when an alternative form of the gene is present on the other chromosome of a pair.

Down's syndrome a congenital condition in which one extra chromosome is present; it results in mental backwardness, short stature and an extra eyelid fold.

endocrinology the branch of science dealing with the structure, function and diseases of the body's ductless glands.

environment the total surroundings of an organism, including both the inanimate world and any other organisms present.

enzymes proteins produced by organisms that act as catalysts for specific chemical reactions within the body.

enzymopathy a disease in which the body is unable to produce one of the normal enzymes, due to a gene defect.

eugenics studies or breeding programs having as their objective the production of an improved, or less diseased, stock.

evolution the process of change by which present-day animals, plants and other organisms have arisen from older, and often simpler, species.

F1 the first generation offspring of a cross between two parents.

F2 second generation offspring, the result of a cross between two of the F1 generation.

Fallopian tubes the tubes down which an egg travels to the womb after its release from the ovary.

fetus a baby within the womb, from the time organs begin to form until birth.

fetoscope a device using fiberoptics which is inserted into the womb to view the fetus.

fiberoptics the use of the property of internal reflection of light in narrow glass fibers to direct light, and receive images, down these fibers (even if they are curved).

fission splitting into two or more, a common method of reproduction in simple organisms.

fixing the incorporation of an element, for example atmospheric nitrogen, into living tissue.

gametes the male and female sex cells that fuse to produce a new individual.

gastrointestinal of the stomach and intestines.

gene therapy attempts to treat disease due to missing or defective genes by insertion of replacement genes into the cells of the body.

genes the units of inheritance that control particular characteristics or capabilities. Genes are located on the chromosomes of the cell nucleus and consist of segments of DNA molecules.

genetic code the sequence of bases in the DNA molecule that ultimately determines the compounds a cell can manufacture and all other physical aspects of the organism, and also forms the basis of heredity.

genetic counseling medical advice which advises on the likelihood of the appearance of possible traits or defects in offspring on the basis of what is known of the genetic makeup of the parents.

genetic engineering artificially making or modifying genes, and inserting them in organisms to serve the purposes of science or industry.

genital herpes a virus disease producing painful small blisters in the genital region.

genotype the genetic constitution of an individual as opposed to the physical appearance.

Graafian follicles the sacs within the ovary containing the developing ova.

growth hormone a hormone from the anterior pituitary gland that influences growth and metabolism.

guanine a base found as one of the constituents of DNA. In the double helix of DNA it is always bonded to cytosine.

hemophilia impaired ability to produce clotting in blood after injury, a characteristic that is a sex-linked inherited trait.

heterozygote an organism in which two different forms of the same gene are present.

homologous pair chromosomes within a cell which form an exactly corresponding pair, each with a set of genes governing the same characteristics.

homozygous with both of a pair of genes identical.

hormones substances produced in one part of the body, strictly a ductless gland, which are conveyed via the bloodstream to another part of the body where they have their effect.

Huntington's chorea a disease involving degeneration of the brain that results in aimless uncontrollable movements. It is hereditary, but the time of its onset varies between patients.

hybridization in genetics, breeding between individuals with different characteristics.

hybrids the results of hybridization.

identical twins twins which have developed from the same fertilized egg.

immune system those parts of the body including, for example, the lymphatic system, that are concerned in the body's reaction to foreign substances.

in utero intrauterine.

in vitro in the test-tube or in laboratory apparatus, rather than in a living organism.

inbreeding breeding within a restricted population among whom individuals may be closely related.

independent assortment the principle that the way that one pair of genes divides between sex cells does not influence the way that a different pair divides. This may not be the case if there is linkage.

inducers enzymes which play a part in switching on the processes by which genes start to manufacture proteins.

insulin a hormone produced by the pancreas that controls the metabolism of glucose.

interferon a substance produced by cells that have been subject to virus attack that acts against other viruses.

intrauterine within the uterus or womb.

inversion a segment of chromosome in which copying has gone wrong at cell division, so that a section of chromosome is reversed.

karyotype the number and shape of chromosomes within a cell nucleus.

Klinefelter's syndrome a condition in which a man has an extra X chromosome in addition to the normal XY pair of sex chromosomes; this usually results in his being tall, sterile, and mentally retarded.

leukemia a form of cancer affecting the white blood cells.

linkage the association of two hereditary characteristics because the genes that code for them are situated close together on the same chromosome.

malignant describes a cancer which invades tissues at locations other than its original site.

mean the total sum of a set of quantities divided by the number in the set; the average.

meiosis the process of cell division by which the sex cells form, so that the number of chromosomes in each cell is only half the number in a normal body cell.

Mendelian adjective describing the "laws" of inheritance, which were first described by Gregor Mendel.

messenger RNA ribonucleic acid which is synthesized in the nucleus and is transported to the ribosomes in the cell taking "instructions" for making proteins.

metabolism the chemical changes taking place within an organism, whether building up or breaking down body substances.

metabolite any substance produced by metabolism.

microflora the collection of bacteria and other small organisms present in an environment.

micrograph a photograph obtained through a microscope.

micropipette a narrow-bore tube which can be used to suck up or inject substances within cells under the microscope.

mitochondria the minute bodies or organelles within a cell which act as the powerhouses for the cell's energy.

mitosis the type of cell division that takes place to produce typical body cells, the two new cells each having the same number of chromosomes as the original.

mongolism Down's syndrome.

monozygous formed from a single egg, as identical twins are.

mRNA messenger RNA.

mutagen an agent which can cause a mutation.

mutation a change in the genetic constitution of an organism.

natural selection the evolutionary principle that the "success" of an organism – whether it survives and breeds – is determined by how well it fits in its environment compared with all of its competitors.

neural tube the tube formed in an early embryo that is destined to produce the central nervous system.

night-blindness inability to distinguish objects normally in very dim light, or at night, often due to vitamin A deficiency.

nits the eggs of a louse.

nondisjunction failure to separate, as can occur abnormally with chromosomes during meiosis.

normal distribution a distribution of quantities symmetrically about a mean figure.

nucleotide one of the single units of which the long chain of a nucleic acid is composed; each nucleotide consists of acid, sugar and base.

oncogenes genes with the capacity to cause cancer.

ontogeny the embryonic development and growth of an individual.

osteoporosis a degenerative disease in which calcium is lost from the bones, leaving them with a porous structure.

parathyroid hormone a hormone, secreted by the four small parathyroid glands in the neck region, that helps control calcium levels in the blood.

pathogens disease-producing microorganisms.

penetrance the percentage frequency with which a gene shows its effects.

phages viruses that attack bacteria.

phagocytes cells that engulf and destroy bacteria.

phenotype the actual characteristics shown by a living organism, as opposed to its genetic makeup.

phenylketonuria a disease of genetic origin in which the enzymes that normally break down the amino acid phenylalanine are missing, causing it to accumulate in the bloodstream and eventually result in brain damage unless treated.

phylogeny the evolutionary history of descent of an organism.

PKU phenylketonuria.

plasmid a small body of DNA that is part of the genetic material of a bacterium.

polygenic inheritance inheritance in which one characteristic is governed by several genes.

porphyria a disease of genetic origin with symptoms that include red urine resulting from an excess of porphyrins in the body.

porphyrins a group of organic pigments, some of which combine with metal atoms to give respiratory pigments such as hemoglobin.

primary structure the sequence of amino acids along a protein molecule as it would appear if the protein was stretched out as a long string. In practice proteins may be folded and show cross-bonding.

probabilistic on the basis of probability rather than certainty.

probability the mathematically calculated likelihood of an event.

progeny the offspring of a mating.

prognosis the probable course of a disease or illness, as forecast by medical opinion.

promoter a site on an RNA molecule that promotes the synthesis of messenger RNA.

proteins large molecules composed of amino acids that are essential components of living things, for example forming many of their enzymes and structures.

recessive describes a hereditary characteristic that appears only when both of a pair of chromosomes have the gene in that same form. If one of the pair has a different, dominant, gene, the effect of the recessive gene is always masked.

recombinant DNA research research which involves working with segments of DNA which are subsequently "recombined" into a larger molecule of DNA.

recombination regrouping of genes caused by crossing-over during meiosis.

repressor a site on a DNA molecule acting to block the production of messenger RNA.

restriction enzyme an enzyme with the property of slicing DNA into regular fragments.

restriction fragment length polymorphism (RFLP) a length of DNA which will combine with a labeled "probe" of a fragment of DNA, and can be used to test for the presence of a particular gene.

ribonucleic acid a long-molecule compound found in all cells which carries instructions from the genes of DNA and regulates the assembly of amino acids into proteins.

ribosomes granular structures within the cell, where amino acids are assembled into proteins.

RNA ribonucleic acid.

RNA polymerase an enzyme that helps assemble RNA molecules.

serum-transmitted disease a disease which can remain infective in blood serum and can thus pose a risk in blood transfusions.

sex chromosomes the chromosomes that determine the sex of an animal. Human females have two X chromosomes, males have one X and a smaller Y chromosome.

sex-linked traits characteristics that derive from a gene on a sex chromosome, and may therefore be shown only by one sex.

sickle-cell anemia a severe anemia in which red blood cells are collapsed to a characteristic sickle shape.

somatic belonging to the ordinary body cells rather than the sex cells.

sonography a technique for obtaining images of an organ or fetus inside the body by means of external scanning using ultrasound.

species a particular kind of animal or plant. All members of the same species are potentially capable of breeding with any other member of the opposite sex.

spina bifida a developmental defect of the lower part of the spinal column.

spore the specialized reproductive cell dispersed by some types of plant and fungus; also, a resting stage of some bacteria.

synthesis the making of a substance.

thalassemia a type of anemia of genetic origin.

thymine a base found as one of the constituents of DNA. In the double helix of DNA it is always bonded to adenine.

thymus gland a ductless gland near the base of the neck, developed in children but regressing in adults.

toxins poisons produced by living organisms.

transfer RNA RNA which attaches amino acids to the ribosomes during protein manufacture.

translocations exchanges, in abnormal cell divisions, of segments of chromosomes belonging to two different pairs.

triplet of bases a sequence of three bases along a DNA molecule that code for a particular amino acid in protein synthesis.

trisomy a condition in which three of a particular chromosome are present in a cell, rather than the normal pair.

tRNA transfer RNA.

ultrasound sound above the range of human hearing.

uric acid a breakdown product of nitrogen metabolism, used in some animals as an excretory product, but in humans liable to produce gout in the joints if present in excess.

vaccine a preparation of a dead or weakened pathogen introduced into the body to stimulate the production of antibodies which give immunity against the live pathogen.

viruses minute organisms that can replicate themselves only within the cells of another living organism. Many produce diseases in their host.

X-linked describes a characteristic for which the gene resides on the X chromosome.

X-ray diffraction a technique in which molecules are bombarded with X rays. The pattern of diffraction of the rays allows the structure of the molecule to be deduced.

Illustration Credits

Introduction
6, *La Nace* by H. Rousseau/Jeu de Paume/Bridgeman Art Library

Ancestral Legacies
8, Zefa/Heilman. 10 *The Harvesters* by Pieter Breughel/Robert Harding Associates. 11, Courtesy of the Trustees of the British Museum 12, Biophoto Associates. 13, J.B. Davidson/ Survival Anglia. 14, Ann Ronan Picture Library 15 top, Mary Evans Picture Library. 15 bottom, Robert Harding Associates. 16, Associated Press. 17, Mary Evans Picture Library. 18 left, The Moravian Museum, Brno. 18 right, **N. Kemball**, 19, **N. Kemball**. 20, Biophoto Associates. 21, Tony Stone Worldwide. 22 left, Dr. G. Schatten/ Science Photo Library. 22 right, Popperfoto. 23, Farmer's Weekly. 24, **Mick Gillah**. 25, *Philip IV* by Rubens/ Uffizi Gallery, Florence/Bridgeman Art Library. 26 left, US Department of Agriculture. 26 right, Biophoto Associates. 27, Nobelstiften, Sweden. 28 left, *Flower Girls at Covent Garden* by A.C. Cooke/O & P Johnson Ltd/ Bridgeman Art Library. 28 right, *Young Lady in a Conservatory* by J.M. Bowkett/ Roy Miles Fine Paintings/Bridgeman Art Library. 29, Novosti Press Agency. 20, **Mick Saunders**. 31, Dr. Alesk/ Science Photo Library.

Inherited Possessions
32, Tony Stone Worldwide, 34 top, Spectrum. 34 bottom, Cambridge Evening News/Scottish Society for the Prevention of Cruelty to Animals. 35, **Mick Saunders**, 36 left, Dr. G. Schatten/Science Photo Library. 36 right, Biophoto Associates. 37, Clement Clarke International. 38, **Mick Saunders**, 39, **Mick Saunders**. 40, Queen Victoria and the Royal Family Boarding the Train to Scotland/ Museum of British Transport/ Bridgeman Art Library. 41 top, **Mick Saunders**. 41 bottom, Mary Evans Picture Library. 42, Godfrey Argent. 43, Heather Angel/Biofotos. 44, John Wiley & Sons, New York. 45, UKAEA Harwell. 46, **Mick Saunders**. 47, Dr. A. Leipins/Science Photo Library. 48, Sally and Richard Greenhill. Foldout, (outside A) John Walsh Science Photo Library. (outside B) National Institutes of Health/Science Photo Library, (inside) **Mick Gillah**, (outside C) National Institutes of Health/Science Photo Library, (outside D) Petit Format/Nestle/Science Photo Library. 49, Courtesy of University of Indiana, Bloomington. 50, Mary Evans Picture Library. 51 left, Tony Stone Worldwide. 51 right, *Mad Kate* by J. Fuseli/Goethe Museum, Frankfurt/ Bridgeman Art Library.

Cracking the Code
52, *The Bridge Family* by John Constable/Tate Gallery/Bridgeman Art Library. 54, Biophoto Associates. 55 left, Biophoto Associates. 55 right, Biophoto Associates. 56, Camera Press. 57 top, **Aziz Khan**. 57 bottom, Biophoto Associates, 58/59 left, Gene Cox. 59 right, **Mick Saunders**. 60, Zefa. 61 top, Sally and Richard Greenhill. 61 bottom, Biophoto Associates. 62, Biophoto Associates. 63, Granada TV Hilda Physick Agency. 64, Biophoto Associates. 65, G. Murti and D. Parker/ Science Photo Library. 66, Popperfoto. 67 **Mick Saunders**. 68 left, Bill Longcore/Science Photo Library. 68 right, Sklar and Perpe/Science Photo Library. 69. **Mick Saunders**. 70, Biophoto Associates. 71, Sally and Richard Greenhill. 72 UKAEA Harwell. 73 left, St. Bartholomew's Hospital, London. 73 right, John Watney.

Inborn or Acquired?
74, Survival Anglia Ltd. 76, C. James Webb. 77, Mary Evans Picture Library. 78, Popperfoto. 79 top, **Mick Gillah**. 79 bottom, Stephen Dalton/Natural History Photographic Agency. 80 top, *Christopher Barker* by kind permission of the trustees. Down House. 80 bottom, Imitor. 81, *Proserpine* by Dante Gabriel Rosetti/Tate Gallery/Bridgeman Art Library. 82, **Mick Saunders**. 83, Tony Stone Worldwide. 84, *Sa Ga Yeath – King of Magug*/Private Collection/ Bridgeman Art Library. 85 left, Godfrey Argent. 85 center, Nobelstiften, Sweden. 85 right, Godfrey Argent. 86, Biophoto Associates. 87, *Sir Walter Raleigh's First Smoke*/Private Collection/ Bridgeman Art Library. 88, Zefa. 89, 90, Mary Evans Picture Library. 91, Popperfoto. 92, Arthur D'Arazien/ Imagebank. 93 top, Bryan and Cherry Alexander. 93 bottom, Mansell collection.

The Gene Machine
94, AT & T, Bell Laboratories. 96, Tony Brain and Elsa Hemming/Science Photo Library. 97 top, Nobelstiften, Sweden. 97 bottom, Nobelstiften, Sweden. 98 top, **Mick Saunders**. 98 bottom, Biophoto Associates. 99, **Mick Saunders**. 100, John Walsh/Science Photo Library. 101, Professor A. Schell/ Max Planck Institute/Bayer U.K. 102, P.A. McTurck, University of Leicester and David Parker/Science Photo Library. 103, Tony Stone Associates. 104 top, Biophoto Associates. 104 bottom, Genentech. 105 left, Genentech. 105 right, *Les Musiciens de L'Orchestre* by Degas/Louvre, Paris/ Bridgeman Art Library. 106, C. James Webb. 107, Muscular Dystrophy Group. 108 top, John Watney. 108 bottom, John Watney. 109, Mary Evans Picture Library. 110, Mr. Kobel & LDU Pasquier, Institute for Immunology, Basel. 111 top left, D. Fowlger/ Rothamsted Experimental Station. 111 middle left, D. Fowlger/Rothamsted Experimental Station. 111 bottom left, D. Fowlger/Rothamsted Experimental Station. 111 right, D. Fowlger/ Rothamsted Experimental Station. 112 top, Monsanto. 112 bottom, Biogen. 113, Genentech.

Screening the Genes
114, *The Well at Gruchy* by J.F. Millet/ Victoria and Albert Museum/ Bridgeman Art Library. 116 top, C. James Webb. 116 bottom, C. James Webb. 117, Mary Evans Picture Library. 118, **Mick Saunders**. 119, **Mick Saunders**. 119 (inset), Sally and Richard Greenhill. 120, **Mick Saunders**. 120 (inset), National Portrait Gallery, London. 121, National Portrait Gallery, London. 122, Mary Evans Picture Library. 123, *Monte Carlo* by Georges Iakoulou/Edimedia. 124 left, Dr. Jack D. Singer. 124 right, Dr. Jack D. Singer. 125, Paul Wilson/Rex Features. 126, **Mick Saunders**. 127, **Mick Saunders**. 128 left, C. James Webb. 128 right, C. James Webb. 129, C. James Webb. 130, Dan McCoy/ Rainbow. 131, G.J. Images/Imagebank. 132, Dr. Jack D. Singer. 133, John Watney.

Changing the Blueprint
134, *Run to the Stars* by John Harris/ Young Artists, London. 136, Punch, London. 137, Zefa. 138, Mary Evans Picture Library. 139, Mary Evans Picture Library. 140, *The Alchemist* by Cornelis Bega/H. Schickman Gallery, NY/Bridgeman Art Library. 141, *The Farmyard* by Vincent Maddesley/Private collection All Rights Reserved/ Bridgeman Art Library. 142, Holt Studios. 143 top left, Ron and Christine Foord. 143 top right, Dr. Jeremy Burgess/Science Photo Library. 143 bottom, Farmer's Weekly. 144 top, **Aziz Khan**. 144 bottom, Tony Stone Worldwide. 145, Tony Stone Worldwide. 146, **Mick Saunders**. 147, **Mick Saunders**. 148, **Mick Saunders**. 149, Jacob Sutton/Reflex. 150, Mary Evans Picture Library. 151, **Mick Saunders**.

Index

Page numbers in bold type indicate location of illustrations